About Island Press

Island Press, a nonprofit organization, publishes, markets, and distributes the most advanced thinking on the conservation of our natural resources—books about soil, land, water, forests, wildlife, and hazardous and toxic wastes. These books are practical tools used by public officials, business and industry leaders, natural resource managers, and concerned citizens working to solve both local and global resource problems.

Founded in 1978, Island Press reorganized in 1984 to meet the increasing demand for substantive books on all resource-related issues. Island Press publishes and distributes under its own imprint and offers these services to other nonprofit organizations.

Support for Island Press is provided by Apple Computers Inc., Mary Reynolds Babcock Foundation, Geraldine R. Dodge Foundation, The Charles Engelhard Foundation, The Ford Foundation, Glen Eagles Foundation, The George Gund Foundation, William and Flora Hewlett Foundation, The Joyce Foundation, The John D. and Catherine T. MacArthur Foundation, The Andrew W. Mellon Foundation, The Joyce Mertz-Gilmore Foundation, The New-Land Foundation, The J. N. Pew, Jr., Charitable Trust, Alida Rockefeller, The Rockefeller Brothers Fund, The Florence and John Schumann Foundation, The Tides Foundation, and individual donors.

About NRDC's Atmosphere Protection Initiative

The Natural Resources Defense Council (NRDC) is a private nonprofit environmental protection organization founded in 1970. NRDC has its principal offices in New York City, Washington, D.C., San Francisco, and Los Angeles. NRDC's staff of lawyers, scientists, and resource specialists address a range of critical environmental problems in the United States and worldwide. NRDC is supported by its more than 130,000 members.

In 1988, NRDC launched its Atmosphere Protection Initiative (API), which is a coordinated, multidisciplinary effort to address the related threats to the integrity of the earth's atmosphere—global warming, ozone depletion, acid rain, and urban smog. Actively participating in the API are more than a dozen NRDC experts on energy conservation, forestry and agriculture, international environment, air pollution control, coastal protection, and nuclear energy.

THE
RISING
TIDE

THE RISING TIDE

Global Warming and World Sea Levels

LYNNE T. EDGERTON

Natural Resources Defense Council

Foreword by George M. Woodwell

UNIVERSITY LIBRARY
GOVERNORS STATE UNIVERSITY
UNIVERSITY PARK, IL. 60466

ISLAND PRESS

Washington, D.C. □ *Covelo, California*

© 1991 Natural Resources Defense Council

All rights reserved. No part of this book may be reproduced in any form or by
any means without permission in writing from the publisher: Island Press, Suite
300, 1718 Connecticut Avenue NW, Washington, D.C. 20009.

Library of Congress Cataloging-in-Publication Data

Edgerton, Lynne T.
The rising tide : global warming and world sea levels / Lynne T.
Edgerton ; foreword by George M. Woodwell.
p. cm.
Includes index.
ISBN 1-55963-068-X. — ISBN 1-55963-067-1 (pbk.)
1. Global warming.   2. Climatic changes.   3. Sea level.
4. Science and state.   5. Science and state—United States.
I. Title.
QC981.8.G56E34   1991
363.73'87—dc20   90-5376
CIP

Printed on recycled, acid-free paper

Manufactured in the United States of America
10 9 8 7 6 5 4 3 2 1

QC 981.8 .G56 E34 1991

Edgerton, Lynne T.

The rising tide

278440

For Brad, whose shared commitment
to a healthy environment has made all the difference;
and for our children, Lauren and Ford,
who shall inherit the Earth

Therefore whoever hears these sayings of Mine, and does them,
I will liken him to a wise man who built his house on the rock,
*   But everyone who hears these sayings of Mine, and does not*
do them, will be like a foolish man who built his house on
the sand.

MATTHEW 7:24–26

CONTENTS

FOREWORD

No one knows how rapidly the earth will warm or how fast the oceans will rise against the land as glaciers melt, as oceanic water warms and expands into the oceanic basins, and as continental plates continue their puzzling gyrations. What we do know is that humans are causing a series of changes in the earth that make the earth less habitable; that the human grip on habitat is weak and weakening daily. A continuously warming earth will not forever support people; a continuously rising sea is the enemy of coastal dwellers everywhere. The progressive submergence of islands, the flooding of low-lying coastal areas around the world, and the larger storms associated with a warmer atmosphere will generate a new wave of refugees among a global population that is adding a million new people every four days to numbers already grown beyond the limits of sustainability. A rising sea is no blessing in such a world.

And yet, political winds blow strongly against the tide of science and elementary commonsense. The same corruption of government and perversion of governmental purpose that have given the United States a decade of deficits, a soaring burden of debt, a bloated military, and a rapidly crumbling public purpose deny the reality of a warming earth and demonstrate a willingness to make war to protect the United States's right to speed the warming. At the recent final meeting of the Intergovernmental Panel on Climatic Change—a meeting that was engaged in a week-long effort to approve a special report on the warming of the earth to the General Assembly of the United Nations—a delegate from Tarawa, a war-seared Pacific atoll, drew applause when he pointed to the imminence of a watery end for his newly-formed nation and urged action to avoid such catastrophe. The applause was universal and included those delegates who had spent the week in officially sanctioned systematic erosion of the sense and purpose of the U.N. study, delegates whose every purpose for days had seemed focused on actions that would complete the destruction started in those bloody days of war forty-six years ago.

Lynne Edgerton and the Natural Resource Defense Council (NRDC) have provided the most comprehensive review yet of the implications of a rapid warming of the earth for those who dwell at sea level around the world. While the subject equally draws flames of passion and concern from scholars who see the hazards to the human condition and from those whose oxen will be gored when the world moves, as it must, to stop the destruction of forests and the further use of fossil fuels, Lynne Edgerton has adopted the dispassionate objectivity that has marked the NRDC's two decades of service in addressing environmental affairs of state.

It is a delight to see this scholarly book, one in a series of penetrating analyses from that special, growing community of scholars supported by conservation interests around the world. These works are aimed specifically at strengthening the ability of governments to function effectively in protecting the public interest in a world in which the issues are more and more complicated, both technically and politically. Ms. Edgerton has done a masterful job.

—George M. Woodwell

PREFACE

One of the most pressing environmental problems of the twenty-first century will be the general warming of the global atmosphere. Warming occurs when various gases are trapped in the atmosphere and block the escape of the earth's radiation to space. Although this "greenhouse effect" is a natural phenomenon, it is widely believed that human actions have caused it to increase at an unprecedented rate. In the past century, heavy industrial use raised the naturally occurring levels of carbon dioxide by more than 25 percent and significantly increased the levels of methane, nitrous oxide, and chlorofluorocarbons. Warmer atmospheric temperatures cause increased atmospheric water vapor, which in turn amplifies the warming of the atmosphere in what has been called an uncontrolled global experiment. Current usage would escalate these trends.

The United States—and the world—can minimize global warming by taking decisive action now. Increased energy efficiency and the development of renewable energy sources can lead the way. Halting the global production and use of harmful chlorofluorocarbons, reducing global carbon dioxide emissions, and stopping deforestation, if accomplished soon, can greatly lessen the risk of catastrophic global climate change and buy time for the development of technological solutions.

We can limit the amount of global warming—but we are probably too late to eliminate it altogether. Some greenhouse gases, such as chlorofluorocarbons and carbon dioxide, have life expectancies in the atmosphere exceeding 100 years. Furthermore, there is no known way to remove them from the atmosphere once the buildup occurs. Thus we must adapt to the likely consequences of warming—at least the most dangerous ones—including the potentially destructive effects of rising sea levels, changed ocean circulation, and increasingly frequent and intense storms on coastal communities and ecosystems.

As a minimum "insurance" policy, coastal areas should now be planning for a significant global sea level rise by the year 2050. Low-lying deltas and barrier islands will be especially vulnerable. Rising sea levels are expected to lead to the loss of coastal ecosystems (including wetlands and estuaries) and coastal protection systems (such as mangroves and coral reefs), as well as coastal barriers, ports, coastal agriculture, and critical habitats.

Fisheries and marine resources vital to endangered marine species will be vulnerable to global change-induced effects—such as shifts in ocean circulation patterns, increased turbidity, water temperature changes, and increased storm activity—and to the possible loss of spawning and nursery areas. The exploitation and conservation of living marine resources are important to both coastal and landlocked states. Moreover, rising seas will cause increased salinization of coastal rivers and groundwater sources and increase the opportunity for release of hazardous substances due to potential flooding of low-lying hazardous waste sites or solid waste disposal facilities, especially from storm surge during major coastal storms.

Although the current consequences of climate change rank low relative to other environmental problems, the future consequences of climate change rank high, especially for densely developed coastal communities and for highly productive coastal ecosystems. Although we cannot completely prevent the adverse effects of climate change on our coasts from occurring, we can minimize the damage to coastal resources and human settlements by initiating a comprehensive program for adapting to the inevitable consequences of sea level rise and global warming.

We can acknowledge that short-term measures—such as directing new development and population away from areas vulnerable to flooding and taking steps to preserve upland buffers to wetlands—can substantially reduce many adverse effects of sea level rise and global warming.

We can recognize that adaptation planning must not be based solely on cost-benefit analyses—which inevitably undervalue important coastal ecosystems, as well as human preferences for stable, safe communities.

We can recognize that there is a time lag of several decades between public recognition of the problem and the actual implementation of coastal protection strategies. U.S. Army Corps of Engineers projects average approximately twenty-six years from conception to completion.

This book outlines state, national, and international actions to respond to the effects of sea level rise and global climate change on coastal communities and ecosystems. The analysis includes:

- Implications: What will sea level rise mean in terms of damage to human health, property, and natural resources?
- Federal and state agencies: Who is responsible for coastal policies? How can existing programs be used to adapt to sea level rise?
- International actions: What will sea level rise mean for other countries? How should we model our international efforts?
- Recommendations: What should the international institutions, the president, Congress, and federal and state governments do now?

The world is finally beginning to wake up to the impending climate change. But few realize that combating warming alone will not be sufficient—we must also prepare ourselves for the consequences of our past actions.

ACKNOWLEDGMENTS

I especially wish to thank John H. Adams, executive director of the NRDC, for his patience and encouragement over the course of this project. I also thank those members of the NRDC staff who edited significant portions of the text: Sarah Chasis and Dan Lashof. Trish Mace made editorial suggestions on Chapter 3. In addition, I thank other members of the NRDC staff who made valuable suggestions, contributed time and expertise, or reviewed the text: Paul Allen, Dick Ayres, Peter Borelli, Ralph Cavanagh, David Doniger, David Goldstein, Jacob Scherr, Lisa Speer, and David Wirth. Lauren Osborne and Marisa Venegas helped to edit an earlier version of the manuscript, and Cathy Dold nurtured the process. A special note of thanks goes to Anne Whalen and CarolAnne Cohen, who patiently and skillfully brought this book through countless revisions.

Drafts of this book were reviewed for technical accuracy and explanation of facts. I am grateful to those who contributed comments and assistance in explicating the complex scientific and legal issues raised by policies responding to sea level rise—especially Stephen Leatherman, Vivian Gornitz, James Tripp, Nick Robinson, Dean Abrahamson, James Broadus, George Woodwell, and James Titus.

Especially helpful were state coastal managers who responded to NRDC questionnaires and reviewed the sections of the book concerning their policies—including James H. Balsille (Florida), Robert Cortright (Oregon), Stephanie A. D'Agostino (New Hampshire), Stephen M. Dickson (North Carolina), Sally S. Davenport (Texas), Peter M. Douglas (California), David S. Hugg III (Delaware), William Johnson (Pennsylvania), Gered Lennon (South Carolina), Jerry L. Louthain (Washington), Jan Mills (Alaska), Jerry Mitchell (Mississippi), David Owens (North Carolina), Rich Shaw (North Carolina), George R. Stafford (New York), Roger A. Ulveling (Hawaii), John R. Weingart (New Jersey), and Christopher F. Zabawa (Maryland).

I believe the tables found in Appendixes D and E will be especially helpful to state coastal zone managers, and I thank Paul Klarin and Marc J. Hershman for permission to reproduce the tables in this book.

Chapter 1

GLOBAL WARMING: FACTS AND IMPLICATIONS

In recent years, activities such as burning fossil fuels, leveling forests, and producing synthetic chemicals such as chlorofluorocarbons (CFCs) have unleashed into the atmosphere large quantities of carbon dioxide (CO_2) and other "greenhouse" gases. These gases are warming the earth at an unprecedented rate. If current trends continue, they are expected to raise the earth's average surface temperature by at least 1.5° to 4.5° C, or more, in the next century—with warming at the poles perhaps two to three times as large as warming at the middle latitudes.[1]

Is It Certain to Occur?

While large uncertainties remain about the timing and ultimate magnitude of climate change, recent research supports the fundamental understanding of the greenhouse effect that has emerged over the last decade. There is little dispute that there has been a significant increase in the global concentrations of heat-trapping gases during the last century. And the overwhelming conclusion from the body of published scientific literature is that this greenhouse gas buildup poses enormous risks.

An unusual aspect of the issue, however, is that the atmospheric warming observed over the past decade does not yet exceed historical natural climate variations, although the five global-average warmest years in the past century occurred in the 1980s. Thus policymakers are

faced with considerable uncertainties. There is worry that current estimates of future warming may be too high, that the uncertainties in
climate computer models may make action premature, that climate-
stabilizing (negative) feedbacks have been omitted from current models.

Even assuming the legitimacy of these concerns, however, the main-
stream scientific view of the greenhouse effect is not based solely on
models. It is supported by observations of other planets, seasonal variations in the earth's climate, and, perhaps most important, on long-term
changes in the earth's climate over the last 160,000 years. This evidence,
in addition to model results, has been reviewed by numerous national
and international scientific bodies—including the National Academy of
Sciences, the Scientific Committee on Problems of the Environment,
NASA's Goddard Institute for Space Studies, NOAA's Geophysical Fluid
Dynamics Laboratory, and various laboratories of the Department of
Energy (DOE). In an August 1990 report, the Intergovernmental Panel
on Climate Change established by the United Nations Environment
Program and the World Meteorological Organization summarized the
scientific facts of which it is certain, as follows:

- There is a natural greenhouse effect that already keeps the earth
  warmer than it would otherwise be.
- Emissions resulting from human activities are substantially increasing the atmospheric concentrations of the greenhouse gases: carbon
  dioxide, methane, chlorofluorocarbons, and nitrous oxide. These
  increases will enhance the greenhouse effect, resulting on average
  in an additional warming of the earth's surface. The main greenhouse gas, water vapor, will increase in response to global warming
  and further enhance it.[2]

With regard to temperature observations, there are two key points.
First, numerous reviews have reaffirmed that after correcting for urban
heat islands and other potential systematic errors, there has been a
statistically significant global warming of about 0.5° C over the last
century. The U.S. temperature record shows less warming than the
global average, but its statistical significance is marginal. This finding
should come as no surprise, for the United States represents only 1.5
percent of the area of the earth and there is no inconsistency between a
global warming and small regions of the earth showing insignificant
warming or even cooling. Second, substantial variability and lags in
global temperatures are expected. Because the oceans have a large heat
capacity and because natural variability and other climate forcings (such
as aerosol loadings) are similar in magnitude to the greenhouse gas

buildup so far, the uneven pattern of global warming observed to date is not at all inconsistent with the climate models.

With regard to feedbacks, these uncertainties are recognized by everyone, but they cut both ways. That is, current predictions are at least as likely to be too low as too high.[3] Indeed, as recently as 1988, scientific predictions were significantly underestimating the magnitude of the ozone hole. Biological and geochemical feedbacks that are not included in the current models are likely to amplify the warming substantially—by perhaps a factor of 2 or more.[4] The hypothesis that warming will be substantially mitigated by increased convection that dries the atmosphere has never been published in the scientific literature; in fact, it has been widely rejected by the scientific community.

The risks associated with global warming are clearly enormous. While there will always be respectable scientists who question global warming predictions, they represent a minority within the scientific community.[5] Although this does not prove that they are wrong, we should not be willing to bet the planet that the National Academy of Sciences, NASA, the Intergovernmental Panel on Climate Change, and other authoritative scientific bodies are wrong and the minority view is right.

There can be no excuse for not pursuing all policy options to reduce greenhouse gases that make economic and environmental sense in their own right. So far the United States is not doing that. Current policies will lead to substantial further accumulation of greenhouse gases. U.S. emissions increased by about 7 percent between 1988 and 1990, and the U.S. share of global emissions, after falling continuously for more than a decade, is now increasing. This is neither necessary nor desirable. This country's carbon dioxide emissions could be reduced 20 percent by the year 2000, increasing our economic competitiveness, reducing our dependence on imported oil, and lowering our contribution to global warming.

How Is Climate Change Measured?

When evaluating the current status of our climate, scientists compare the historical climate record with current observations and with the predictions of climate models. Since a certain amount of statistical variability in world climate is known to exist, they attempt to separate this natural variability, or "noise," from the true signals that represent real changes in the system balance rather than random fluctuations. Such work is carried out by developing "signal-to-noise ratios." The larger the apparent signal compared to the noise, the more certain scientists can be that the global climate is changing.

The key to understanding climate fluctuations often lies in our analysis of these shifts in both the human and proxy records (fossil records). Human records have been useful in elucidating short-term trends, although they are generally considered less reliable if collected before the year 1900. Proxy records are likewise inaccurate, and often incomplete, offering only limited resolution on the scale of hundreds to thousands of years.[6]

On the other hand, ancient natural records provide us with a broad view of our climate's variability over time and indicate the effects of perturbations to the global climate system. By a variety of methods—including the analysis of gases trapped in air pockets of ancient ice flows and the analysis of oxygen and carbon isotopes from deep-sea sediment cores—scientists have been able to construct a historical record of atmospheric composition and global climates. They have found that global temperatures have varied widely throughout time and that sea levels have varied concurrently. Past climate changes, they hypothesize, have historically ensued from the combination of (1) periodic variations in the earth's orbit and changes in solar output that influence the distribution and amount of energy the earth receives from the sun, (2) catastrophic events such as meteorite impacts, volcanic eruptions, or changes in oceanic circulation, and (3) ongoing biological processes. Recent measurements indicate that humans have the ability to alter climate significantly as well.[7]

By comparing historical data obtained from air bubbles trapped in ancient ice with current observations, changes above normal fluctuations in the ratios of certain atmospheric gases have been documented. Among these gases are carbon dioxide, methane, and nitrous oxide, which until the last few centuries had existed in a stable natural balance for millennia due to biological and chemical processes. Over geological time, these major gases have played a vital part in regulating the earth's climate.

The Greenhouse Effect

Water vapor, carbon dioxide, methane, nitrous oxide, and ozone comprise the natural fraction of what are called the greenhouse gases. These gases delay the escape of infrared radiation from the earth into space, thus causing a general climatic warming known as the greenhouse effect (by analogy to the warming that occurs in a greenhouse). Scientists have stressed that this is a natural process—indeed, the earth would be 33° C (63° F) cooler than it is presently if the greenhouse effect did not exist.[8]

Human activities are now rapidly intensifying this natural greenhouse effect. Carbon dioxide concentrations have increased 25 percent since preindustrial times, methane concentrations have doubled, and chlorofluorocarbons have been introduced into the atmosphere for the first time. Human activities have not only directly added these greenhouse gases to the atmosphere, but they have greatly altered the chemistry of the global atmosphere by increasing the production of tropospheric ozone and decreasing the natural destruction of methane, thus further enhancing the greenhouse effect. The rate at which the greenhouse effect is intensifying—the rate of climate "forcing"—is now more than five times what it was during the last century.[9]

During the 1980s carbon dioxide accounted for approximately 55 percent of the climate forcing, while CFCs accounted for about 24 percent. Methane and nitrous oxide together contributed approximately 21 percent. We turn now to the empirical evidence and its implications.

THE INCREASE IN CARBON DIOXIDE

The increase in atmospheric carbon dioxide is due primarily to fossil fuel combustion (coal, oil, and gas) and deforestation. The most precise measurements of atmospheric CO_2 were begun at Mauna Loa, Hawaii, in 1958. (See Figure 1.1.) The rate of observed CO_2 increase since that time has closely followed fossil fuel consumption rates; consequently, the strong implication is that this source dominates any natural source and overwhelms any natural processes of removal.[10]

Global deforestation is believed to be another important contributor to the increasing concentration of carbon dioxide.[11] As land is cleared or burned for agriculture, or harvested for timber, CO_2 is released into the atmosphere from the decay and oxidation of the carbon that had been incorporated in the trees and soil organic matter. These processes are estimated to contribute 10 to 30 percent of anthropogenic CO_2 emissions. Although the future contribution of deforestation or reforestation to atmospheric CO_2 levels could be significant, these aspects of the global carbon budget are little understood and have thus far been inadequately modeled. Control of future CO_2 concentrations would benefit substantially from strengthened research in this area.

If nothing else in the earth's climatic system changed, a doubling of

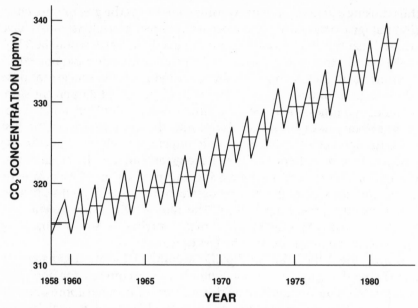

Figure 1.1 *Mean Monthly CO$_2$ Concentration at Mauna Loa, Hawaii*

SOURCE: *National Research Council (1987).*

CO$_2$ would raise the global average temperature 1.2° C.[12] The earth's climatic system, however, is highly complex: A warming from increasing CO$_2$, for example, causes corresponding changes in other parts of the system, such as increasing atmospheric water vapor and altering precipitation and cloud cover. Since the impact of these feedback mechanisms cannot be precisely determined, total warming is difficult to estimate. Sophisticated "general circulation models" (GCMs) of the earth's climate suggest that some of these feedbacks could be quite strong and would raise the warming by several degrees above that calculated by considering the CO$_2$ increase alone. This finding is supported by the empirical evidence from historical changes in CO$_2$ concentrations and climate revealed by the ice core measurements noted earlier.

THE INCREASE IN CHLOROFLUOROCARBONS

Chlorofluorocarbons are produced exclusively in industrial activities. Because they remain in the atmosphere for approximately 100 years before breaking down,[13] CFCs contribute substantially to global warming. Although CFCs comprise a relatively small proportion of the greenhouse gases by volume, they also destroy stratospheric ozone and produce 10,000 to 20,000 times more forcing per molecule in the atmosphere than CO_2.[14] CFCs are produced for such diverse uses as refrigeration and air conditioning, semiconductor manufacturing, degreasing solvents, and foam insulation.

Apart from their warming effects, CFCs have been identified as a major cause of the destruction of stratospheric ozone. As the stratospheric ozone layer thins, human beings become more vulnerable to the adverse health impacts of increased ultraviolet radiation. These impacts include skin cancer, cataracts and other vision disorders, broad systemic disruption of human immune responses, and possibly many other illnesses.[15]

In September 1987, thirty-one nations met in Montreal and reached an international agreement to require a 50 percent reduction in the consumption of five CFCs by the end of this century. The EPA, however, estimates that an immediate 85 percent reduction of CFC emissions would be necessary merely to stabilize the amount of ozone-depleting substances reaching the upper atmosphere. In May 1988, just one day after the U.S. Senate's 83–0 ratification of the Montreal accord, the National Aeronautics and Space Administration (NASA) issued a report demonstrating that stratospheric ozone depletion is dramatically worse than was thought as recently as late 1987. The NASA report documents an unexpectedly rapid thinning of the stratospheric ozone shield all over the globe—with CFCs the likely cause. These alarming findings have added new urgency to NRDC's drive for a total phaseout of CFCs and other ozone-depleting chemicals.

According to the NASA report, even after natural factors are accounted for, satellite and ground-based monitors show ozone losses since 1969 as high as 3 percent over the heavily populated regions of North America and Europe and 5 percent over parts of the Southern Hemisphere. What is more, depletion is occurring at two to three times the rate predicted by the computer models scientists have relied on.

THE INCREASE IN OTHER GREENHOUSE GASES

The growth of atmospheric methane and nitrous oxide has also caused global warming projections to rise. Methane, the major component of natural gas, is released to the atmosphere primarily from decomposition in wetlands, landfills, bogs, and rice paddies. It is also a by-product of digestion in cattle, sheep, and termites. Abiotic sources of methane include leaks from natural gas production and distribution, release of coal-seam gas during mining and processing, and incomplete combustion in slash-and-burn agriculture. Humans are responsible for the 11 percent increase in methane in the 1980s because they grow more rice, raise more cattle, produce more fuel, and clear more land (thus burning more material and possibly enhancing the habitat and food supply of termites). The cause of the continuing increase in the concentration of nitrous oxide is not clear at this time, but it could be due to the expanded use of nitrogen fertilizers around the world to improve agricultural productivity.

Another significant factor in global warming is ozone, which is created in the lower atmosphere (troposphere) by complex photochemical reactions involving nitrogen oxides, methane and other hydrocarbons, and carbon monoxide—the principal components of urban smog.[16] Ozone is not uniformly distributed in the atmosphere like the other greenhouse gases, however, and its climatic impact depends strongly on its altitude. Thus while ground-level ozone is increasing, stratospheric (or upper atmospheric) ozone is decreasing. The overall impact on climate at the present time is unclear.

How Much Warming Is Inescapable?

Are we already experiencing the beginnings of the warming effects of increased concentrations of greenhouse gases? Expected global changes include warmer weather, warmer ocean water, a cooler stratosphere, increased precipitation, and a decrease in the daily range of temperatures.[17]

The evidence for warmer weather is clear. A number of studies have shown that the earth has warmed by about .53° C (1° F) during the last century. Climate records for the 134 years prior to 1986 from the University of East Anglia in Great Britain show that a marked warming trend began around 1890. The three warmest years recorded during that period were 1980, 1981, and 1983—and five of the nine warmest years occurred since 1978. Yet 1987 and 1988 were hotter still. The 1980s finished with the fifth hottest year on record, making it the hottest decade of all.[18]

Moreover, frozen ground beneath the Arctic has warmed 2–4° C (4° to 7° F) over the past century.[19] This rise is consistent with scientists' expectations that the temperature rise from global warming in the polar regions will be several times greater than the global average. The warming of the permafrost was detected through an analysis of two dozen oil exploration wells drilled in Alaska's North Slope since the 1950s.[20]

Despite these indications of global warming, there are no scientifically accepted methods of determining whether or not this warming is due to the greenhouse effect. The climate forcing from increases in greenhouse gas concentrations is only now beginning to emerge above the noise from other factors that influence climate. It may be more than a decade before the warming pattern can be clearly attributed to human activities. Even then, not every year will be warm; the greenhouse effect only increases the *average* temperature and the *probability* of very warm years.

Action to respond to this global warming cannot await scientific confirmation that greenhouse gas increases have already changed the climate. According to model estimates, the observed and postulated increases in the trace gases between 1850 and 1980 have committed the planet to an equilibrium warming of 0.7° to 2.0° C. (Due to the thermal inertia of the oceans, however, we can expect only one-third to one-half of the equilibrium warming to show up in the current observed records.) Thus even if we were to cease all anthropogenic production of greenhouse gases, the global average temperature could rise by about 2.0° C.[21]

The most complete models predict that an increase in greenhouse gases equivalent to a doubling of atmospheric CO_2 would cause a global warming of 2° to 5° C.[22] If current trends continue, the earth could be committed to a warming of this magnitude as early as the year 2020. In Washington, D.C., for example, this could increase the number of days on which the maximum daily temperature exceeds 100° F from one day per year to twelve. In Omaha, Nebraska, that number increases from three to twenty. The number of days with maximum temperature exceeding 90° F increases from about thirty-five days to eighty-five days in both cities.[23] Thus a warming will substantially modify the human environment in the middle latitudes and will have major impacts on the quality of everyday life. But the indirect effects of climate change—such as sea level rise, drought, and forest dieback—may be even more significant than the direct impact of higher temperatures.

Although significant climate change is at this point inevitable, the rate and magnitude of global warming can be limited if bold policies are implemented immediately. Slowing the rate of warming is particularly important to ease the temperature-induced stress on the environment and allow humans time to prepare for the accompanying rise in sea level. A recent study by the World Resources Institute (WRI) suggests four possible scenarios of future energy use and greenhouse gas emissions, in each case estimating the rate of CO_2 increase and associated temperature rise.

WRI's *base-case scenario* assumes no change in current trends of population growth and energy use. It allows for limited technological improvements to increase energy efficiency and minimal development of alternative power sources. In this case, the concentration of CO_2 would reach double preindustrial levels in the year 2060.

The *high-emissions scenario* projects accelerated growth in energy use and no policies to limit CO_2 output. Humans develop few alternative fuel sources, and deforestation is rapid. This scenario predicts an almost exponential growth of atmospheric CO_2 levels, resulting in a doubling by the year 2030.

The *modest-policies scenario* assumes a strong stimulus for improved end-use efficiency and a modest stimulus for the development of non-CO_2-producing energy sources such as solar energy. Governments encourage substantial reforestation efforts and levy taxes on fossil fuels to discourage their use. The CO_2 levels in this case rise at a slow but steady rate to a doubling in the year 2090.

The *slow-buildup scenario* envisions a strong emphasis on improving energy efficiency coupled with the rapid introduction of solar energy. A major global effort focuses on reforestation and ecosystem protection, and high taxes are placed on fossil fuels to encourage fuel switching. The growth of atmospheric CO_2 is quite slow in this case, and total greenhouse forcing could be stabilized toward the end of the twenty-first century.

Thus, according to WRI, a CO_2-associated warming of 1.5° to 4.5° C could be delayed by more than seventy years should strong measures be implemented soon. On the other hand, "if 30 years of delay [are] allowed for removing scientific uncertainties, identifying options, establishing international consensus, and implementing appropriate policies, Earth will be committed to a warming 0.25° to 0.8° C higher than that which would occur if the policies envisioned in the slow-buildup scenario were implemented today."[24]

Strategies for Limiting the Warming

To slow the process of warming, we must implement policies to reduce and limit emissions of the most potent greenhouse gases: chlorofluorocarbons, carbon dioxide, and nitrous oxides. NRDC believes we must begin now to:

- Dramatically reduce net CO_2 emissions by using energy much more efficiently, moving away from fossil fuels in favor of renewable energy sources, and stopping global deforestation and starting massive tree planting
- Phase out harmful CFCs and related gases
- Reduce production and emissions of methane and nitrous oxide

We have no time to waste. The June 1988 Toronto Conference on the Changing Atmosphere recommended that the world reduce its CO_2 emissions 20 percent by the year 2005—just fifteen years from this writing. The Advisory Group for Greenhouse Gases, an international group of distinguished scientists and policymakers, assessed the feasibility of greenhouse gas reductions during four years of workshops and analysis. Their October 1990 draft report concluded that "given aggressive implementation of policies, annual carbon dioxide emissions could be reduced by 25 percent from their 1986 levels by 2005 and 40 percent by 2020. A 20–30 percent reduction in overall present methane emissions also appeared feasible using existing technologies."[25] The United States and other industrialized nations must take the lead. NRDC recommends that the United States begin to take action against global warming by reducing its CO_2 emissions at least 20 percent below 1987 levels by the year 2000.

The necessary actions are cost-effective measures that have many other environmental and economic benefits:

- Eliminating CFCs will prevent another global threat—depletion of the stratospheric ozone layer.
- Using energy more efficiently will help to prevent acid rain and urban smog, increase our international competitiveness, and save U.S. consumers tens to hundreds of billions of dollars each year.

- Preventing the destruction of forests and planting trees will conserve the earth's precious animal and plant species, prevent soil erosion, and protect water supplies.
- Slowing atmospheric warming will delay accelerated sea level rise.

What specific actions must the United States take to achieve the goal of reducing CO_2 emissions 20 percent by the year 2000? A substantial increase in efficiency in each of the following areas would bring about the desired reduction:

Measure	CO_2 Reduction
Higher federal vehicle fuel standards	4.4%
Federal actions to improve lighting efficiency	2.4%
Higher federal appliance efficiency standards	2.5%
Federal actions to promote industrial efficiency	2.8%
State building efficiency standards	2.3%
Actions to promote renewable energy sources	2.8%
Federal conservation reserve forestry program	1.4%
Improved management of forest lands	1.9%
Urban tree planting	1.2%
Improved mass transit	0.3%
Total	22.9%

To accomplish this goal, the federal government should follow the lead of many states and adopt an energy planning process that treats energy conservation as a resource and considers all the economic and environmental costs associated with each supply and conservation option. This least-cost method can enable our nation to identify many other cost-effective conservation measures and guide relevant federal decisions, such as those concerning research and development.

Finally, the federal government should begin at once to reduce our dependency on fossil fuels. One incentive could be a fee on fuels proportional to their carbon emissions. Such a fee would encourage consumers to abandon fossil fuel use while bringing in much-needed revenue to reduce the deficit and provide capital to research and implement new energy programs. Eventually, these new programs would save energy consumers tens to hundreds of billions of dollars each year.

The government should also phase out its massive program to demonstrate new methods of burning coal with fewer emissions of the pollutants that cause acid rain. The so-called "Clean Coal" program now accounts for some 50 percent of the federal energy research and development budget, yet its success would only increase the global warming problem.

Reducing U.S. emissions of CO_2 by 20 percent is vital, but such a reduction alone will not minimize climate change sufficiently. Although the United States now accounts for about 23 percent of global CO_2 emissions, that proportion is dropping. Without major reductions in CO_2 from other parts of the world, we cannot solve the problem. Many Western industrialized nations are already persuaded that substantial reductions of their CO_2 emissions are essential. In October 1990, the European Community reached a compromise agreement to stabilize the twelve nations' joint carbon emissions by 2000 by freezing overall joint carbon emissions at 1990 levels. Joint commitments to reduce CO_2 emissions from these nations would have a major impact on both global emissions and international negotiations. Steps must be taken to ensure that the Soviet Union and the Eastern European nations, which are extremely inefficient in their use of energy, join in cutting emissions as soon as possible.

Furthermore, the major developing countries must join in efforts to minimize CO_2 emissions. Their emissions now account for only about 30 percent of the world total, but in view of their fast-growing human populations and industrialization they are likely to account for 50 percent or more by the year 2025. The United States should use scientific evidence, diplomacy, and offers of assistance to persuade developing nations to join the effort to minimize global warming. These efforts will not be credible, of course, unless we and other industrialized nations have shown that we are willing to do our part.

We should use our bilateral foreign aid program and the World Bank and other multilateral aid agencies to help developing countries take the needed actions to adjust while maintaining growth. We should use the same means to stimulate actions to reduce tropical deforestation—now thought to account for 10 to 30 percent of global CO_2 emissions—and promote tree planting. Bold forestry actions could reverse this situation and make the forests a net sink for CO_2.

It would be especially useful for policymakers to establish long-term goals, or targets, for the rates and magnitude of temperature range and sea-level rise and for the greenhouse gas concentrations in the atmosphere. The Advisory Group on Greenhouse Gases has proposed a temperature rise target that is based primarily on adaptation rates of ecosystems, on a maximum rate of increase of temperature of $0.1°$ C per decade, and a maximum temperature increase of 1.0° C or 2.0° C above preindustrial global mean temperature.[26]

But CO_2 emissions are not the only culprit in global warming. We also must act against harmful chlorofluorocarbons and their chemical relatives, which not only cause global warming but also deplete the ozone layer. Even under the recently negotiated Montreal Protocol, the concentration of these chemicals in the stratosphere will double before the year 2050. They must be phased out as rapidly as substitutes become available. The United States should act domestically to achieve a ban on production and use of CFCs by 1995 and encourage international agreements to achieve the same goal worldwide.

A global treaty is needed to establish a greenhouse gas reduction goal and allocate responsibilities among nations. The treaty should also require a commitment from wealthier countries for increased research into non-CO_2 energy supply technologies and aid for poorer countries working to meet the treaty requirements. Funds might be generated by a worldwide greenhouse gas emissions fee. The United Nations Environment Program should be the forum for negotiation of the agreement. The United States has offered to host a meeting in February 1991 to start the international negotiations for such a convention on the climate. The United Nations Conference on Environment and Development, scheduled for 1992 in Brazil, will provide a potential forum for agreement on such a treaty.

Adapting to the Inevitable

Obviously, the best solution to the problem of global warming would be for the world to adopt the recommendations outlined here. But even if these measures were adopted, greenhouse gas concentrations would continue to rise for several decades. It is not realistic to expect climate forcing to be stabilized immediately. Moreover, some global warming is already inevitable.

What, then, must we do? We must adapt to the warming already in progress and prepare for its future effects. Perhaps the most certain effect of global warming is the accompanying rise in sea level. This rise will result in increased storm damage, pollution, and subsidence of coastal lands—and, unfortunately, it is already under way.

Chapter 2

PLANNING FOR THE FUTURE

There is consensus among scientists that global sea levels have been rising steadily for the past 100 years and that global warming is likely to accelerate the rate of this rise over the next century. Nations should begin now to adapt land uses in order to avoid the worst effects of sea level rise and global climate change on coastal communities and ecosystems.

Because there is substantial uncertainty about the way the ocean and atmosphere interact, scientists almost uniformly present their estimates of future sea level rise in terms of a range of possibilities. This scientific practice poses some conceptual difficulties for coastal planners, but it need not. Virtually all of the principal scientists working in the field believe that the time has come for coastal planning with regard to accelerated sea level rise. Scientists believe that a significantly accelerated global sea level rise is probable during our lifetimes. The planners' task, then, is to work out—as a policy matter—a reasonable and prudent general planning assumption for how much the sea will rise and by when. This chapter reviews both the scientific predictions and the relevant policy considerations. Based on the best scientific information available, the general planning assumption should be for a 0.7-meter global sea level rise by the year 2050.

Past Changes in Sea Level

During the past 35,000 years, sea levels have dropped and risen markedly as global temperatures warmed and glaciers melted. Marine

17

fossils have been found hundreds of miles inland in the United States, indicating that much of this country was once covered by an epicontinental sea. Sea level can be thought of as the dipstick of climate change.

The relative sea level rise at a locality is the sum of effects of global sea level rise combined with vertical land motions due to tectonics, glacial loading, or fluid withdrawal (gas, oil, water). Relative sea level changes are local in nature, involving transient climate phenomena, continental uplift, or subsidence. Louisiana is experiencing a relative rise in sea level because its coasts are sinking relative to the Gulf of Mexico. Scandinavia and Alaska, on the other hand, are experiencing a relative drop in sea level due to uplifting of continental crust in those areas. Temporary climate changes resulted in a relative sea level rise along the western coast of South America for several months during the El Niño event of 1983.[1]

Global sea level changes, by contrast, are alterations in oceanic water levels relative to the earth's gravitational level surface (which is assumed to remain constant). Global rises are primarily caused by a melting of land-based glacier ice and the thermal expansion of ocean water. Global declines in sea level can be caused by increased storage of water on land. Global sea level changes, which occur worldwide, have been fairly common throughout the earth's history.

Current Changes in Sea Level

Global mean sea levels have been rising for the past century. Analysis of 100 years of written records from tide gauges around the world reveals that the global sea level has been rising at a rate of 1 to 2 millimeters annually. Although this trend of sea level rise has been evident for some time, research suggests that the most likely explanation for the accelerating rate of increase is a global increase in temperature.[2]

Three independent lines of evidence corroborate that global mean sea level has been rising during the past 100 years: tide gauge records; erosion of 70 percent of the world's sandy coasts and 90 percent of America's sandy beaches; and the melting and retreat of mountain glaciers.[3] The correspondence between the two curves of rising global temperatures and rising sea levels during the last century appears to be more than coincidental.

Future Changes in Sea Level

There is consensus among scientists that global warming will cause the sea level to rise in two ways: thermal expansion of seawater and melting ice. Upper layers of the ocean are expected to expand since the density of water decreases upon heating. The total expansion depends on the rate of global warming and the way in which heat is mixed below the surface of the oceans. In calculating the total contribution of melting ice to sea level rise, scientists add the separate effects of polar glaciers, alpine or mountain glaciers, and the possible breakup of the Greenland and Antarctic ice caps. (See Table 2.1.)

In 1983, the National Academy of Sciences (NAS) Climate Research Board and the Environmental Protection Agency (EPA) developed a series of scenarios to project sea level rise based on various warming assumptions and ocean heat absorption assumptions. The EPA estimated a global sea rise of 0.26 to 0.39 meter by the year 2025, with a likely 0.91 to 1.37-meter rise by 2075. The NAS estimated a 0.70-meter rise by 2080; however, it did not include any contribution from the melting of Antarctic glaciers. The NAS Polar Research Board subsequently estimated that although the contribution of Antarctica by the year 2100 is most likely to be about 0.30 meter, a 1 to 2-meter contribu-

TABLE 2.1 Contributions to Future Sea Level Rise in the Year 2100 (in centimeters)

Study	Thermal Expansion	Alpine Glaciers	Greenland	Antartica	Total
Hoffman et al. (1986)	28–83	12–37	6–27	12–220	57–368
Thomas (1985)	—	—	—	0–220	—
Hoffman et al. (1983)	28–115	b	b	b	56–345
NRC (1983)	—	10–30	10–30	−10–+100	—
Revelle (1983)[a]	30	12	12	c	70

[a] Contributions in the year 2085.
[b] Hoffman et al. assumed that the glacial contribution would be one to two times the contribution of thermal expansion.
[c] Revelle attributes 16cm to other factors.

SOURCE: *Natural Research Council (1987).*

tion is possible. The first study to incorporate this new information was completed by Hoffman et al., who estimated the rise by the year 2025 to be between 0.10 and 0.21 meter (and 0.36 to 1.91 meters by 2075).[4]

An EPA report to Congress in October 1989 found that global warming could cause a global sea level rise of 0.5 to 2 meters by the year 2100.[5] A 1987 National Research Council (NRC) report estimated that the global sea level rise would be 0.5 to 1.5 meters by 2100.[6] Although sea level could ultimately rise by 6 meters if the West Antarctic ice sheet collapses, such a collapse and melting is expected to occur over a period of a few centuries, if at all.[7]

In 1989, Dr. Mark Meier predicted that while the rate of sea level rise will probably accelerate due to thermal expansion and glacial melting, there may be some mitigating factors influencing the process. These mitigating factors include increased precipitation over Antarctica and Greenland leading to greater storage of water on the land; percolation and refreezing of meltwater from polar glaciers and ice caps prior to running off into the ocean; and the slow dynamic response of ice sheets to warming, which slows the movement of water from the land to the oceans.[8] Taking these factors into account, Meier estimates that sea level rise will be 0.3 meter, plus or minus 0.4 meter, by the year 2050. This estimate is consistent with the preliminary value of expected sea level rise of 0.25 to 0.40 meter by 2050, adopted by the Intergovernmental Panel on Climatic Change (IPCC) for use in assessing adaptive responses.[9] Recently collected data support Meier's hypothesis. Satellite images have revealed that ice has been accumulating over southern Greenland since 1978.[10] No records of recent trends in ice accumulation over Antarctica or northern Greenland have been made as yet.

There are major uncertainties, however, in estimates of future sea level rise. For example, although it has been reported that as much as 20 to 50 percent of the water responsible for the sea level rise experienced during the last century has been due to small glacial melting,[11] other scientists disagree, stating that we do not reliably know the source of water or understand the processes contributing to the rise.[12] If Greenland and Antarctica are indeed accumulating ice overall, then it is that much more difficult to explain the currently observed rate of sea level rise. Our lack of understanding of the mechanisms contributing to relatively recent rises in sea level underscores the uncertainty in predicting future changes.

Furthermore, different outlooks for climatic warming dramatically affect estimates. Indeed Meier, whose estimate is based on a 2–3° C warming by 2050, has stated that if the warming is as high as 4.5° C, the "sea levels may well rise as much as a meter by 2050."[13] Moreover, Meier's estimate of future sea level rise involves only simple physics. It is based on the fact that as the air warms it can hold more moisture and, consequently, produce greater precipitation. Nevertheless, such factors as topography and local wind circulation patterns may be more important than atmospheric moisture in determining future precipitation rates over Antarctica and Greenland.[14] General circulation models, which could be used to evaluate Meier's prediction, function particularly poorly over ice sheets and thus have not been used to test his hypothesis.

Because there is great uncertainty about the interactions of the oceans, atmosphere, and warming, estimates of sea rise will undergo continual revision and refinement over the next century. Better understanding of the melting processes for the Antarctic ice cap and the Greenland ice sheet will develop gradually.

Planning Guidelines

Given this uncertainty, general planning guidelines are difficult to formulate. The adverse effects of sea level rise and the costs of adaptive options may occur at different times. Although discounting may be an appropriate tool for comparing future and present monetary values, it is highly problematic when intergenerational issues are at stake or environmental values such as biological diversity are involved. Moreover, existing generations should be investing in sustainable development. Long lead times are required for land use adaptation, for coastal ecosystem conservation, and for major infrastructure commitments. In most cases, choosing conservative assumptions and planning for a larger rise than actually occurs will be much less costly than planning for a smaller rise than materializes.

When environmental values are at stake or human settlements are at risk, a planning assumption should take into account the fact that the effectiveness of most planning options for reducing the adverse effects of sea level rise and global warming decreases over time. (See Figures 2.1 and 2.2.) Thus the mean or "best" estimate of sea level rise is not appropriate because it does not take the higher risks of underestimating the rise into account. A reasonable guide to policy is to choose a sea level rise assumption that is judged to have only a 10 to 15 percent chance of being exceeded.

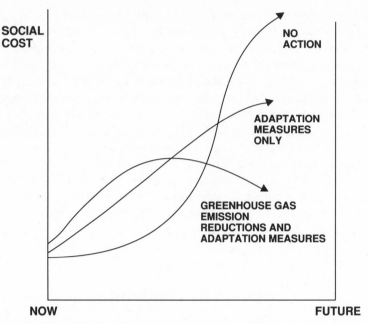

Figure 2.1 *Planning for Sea Level Rise: Tentative Cost Scenarios for Limitation, Adaptation, and No-Action Strategies*

SOURCE: *Adapted from Vellinga and Leatherman (1990).*

The standard deviation is a measure of the range of uncertainty associated with a mean or "best" estimate. In the case of Meier's estimate of sea level rise, there is roughly a probability of one-sixth that the actual sea level rise will be more than 0.7 meter, which is one standard deviation (0.4 meter) above his mean estimate (0.3 meter). Consequently, coastal zone land-use planners should adopt a 0.7-meter rise by the year 2050 as their general planning assumption.

Whenever possible, the uncertainty in future sea level rise estimates should be explicitly considered in the decision-making process. As scientific understanding improves, the specific planning assumption should be modified based on assessments of the Intergovernmental Panel on Climatic Change, the Environmental Protection Agency, and other relevant scientific panels. The general criterion of planning for the mean estimated sea level rise plus one standard deviation will remain valid.

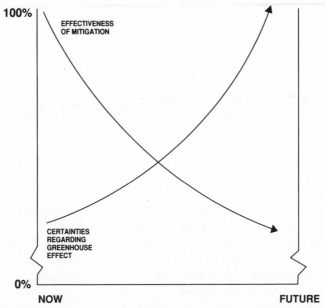

100%

**EFFECTIVENESS
OF MITIGATION**

**CERTAINTIES
REGARDING
GREENHOUSE
EFFECT**

0%

NOW

FUTURE

Figure 2.2 *Impact of Delay on Effectiveness of Mitigation Measures*

SOURCE: *Adapted from Vellinga and Leatherman (1990).*

These estimates of sea level rise are consistent both with recent published scientific data on the subject and with international coastal planning recommendations. Vellinga and Leatherman have persuasively argued that nations should be planning for a 0.5-meter to 1-meter sea level rise by the year 2100.[15] Finally, the effects of sea level rise from global warming must be combined with local conditions to determine the impact on a particular community. Many communities have been studied, and a summary of these impacts is found in Chapter 5.

~~~~~~~~~~~~~~~~~~~~~~~~~~~~~~~~~

Chapter 3

THE MAJOR EFFECTS

Increased global temperatures will lead to a global sea level rise at a rate unprecedented in human history. Temperature and sea level changes will be accompanied by changes in salinity levels, in tidal flooding, in oceanic currents and upwelling patterns, in storm intensities and frequencies, in biological processes, in runoff and landmass erosion patterns, and in saltwater intrusion.

As global sea level rises, more areas of the coast will be below sea level. If the slope of the coast is gentle, a small rise in sea level can result in a considerable submergence of coastal land. On a gentle slope (20:1, horizontal to vertical), ignoring subsidence and uplifting, a 1-meter rise would move the tide line landward by 20 meters (65.6 feet).

The principal environmental effects of sea level rise and climate change will be the loss or alteration of important ecological resources like beaches, barrier islands, and wetlands, as well as increased saltwater intrusion into the nation's estuaries, rivers, and groundwater. Warmer temperatures could also stress certain marine species. Moreover, increased storm intensities and frequencies could cause serious property damage to coastal structures, as well as human injury and death.

Loss of Ecological Resources

BEACHES AND BARRIER ISLANDS

Shorelines in their natural state are dynamic systems that continually advance and retreat with the forces of wind and water. Barrier islands, beaches, and coastal wetlands are all elements of these systems. To

24

understand how sea level rise will affect these resources, one must first understand how the coastal system works.

Beaches are simply lines of sand grains waiting to go somewhere else. They change shape with the seasons, moving down the coast in the direction of the prevailing current, opening and closing inlets and lagoons. Similarly, dunes and islands move landward in response to the sea's assault, changing in a few years or a few seasons, sometimes rapidly enough to watch. Many people go to the same beach summer after summer but do not realize that it is not the same sand or the same dune.

Sand transport is cyclical. In winter, waves pull the finer sand particles out to sea, where they form sandbars. The sandbar protects the land by absorbing some of the energy of strong winter waves. In summer, the gentler seas push fine sand back onto the beach. In this way, a beach is replenished in the summer of sand lost in the winter. A net loss of sand can result if storms carry sand too far offshore or if an above-average stormy season moves more sand offshore than mild waves can bring back.

Sand also moves with currents along the coast. This sand movement creates conflict among owners of contiguous shoreline when the up-current owner builds erosion control structures such as groins to catch the sand, thereby depriving the down-current owner of his ordinary sand supply.

One of the most important direct physical effects of sea level rise is on the East Coast's already threatened barrier island and beach system. Current rates of sea level rise (1 to 2 millimeters per year) already produce significant coastal erosion. Two major factors contribute to the erosion. First, deeper coastal waters enhance wave generation, thus increasing their potential for overtopping barrier islands. Second, shorelines and beaches will attempt to establish new equilibrium positions according to the Bruun rule; these adjustments will include a recession of shoreline and a decrease in shore slope.[1] Figure 3.1 shows how the Bruun rule works.

In the past century, the sea level along the Atlantic coast has risen about 1 foot. Erosion and inundation from a 1-foot rise in sea level can cause a shift of 5 to 100 feet or more landward of the beach depending on the slope of the shore.[2] When the sea rises, barrier beaches and islands will either migrate landward by rolling back over themselves or drown in place. With continual migration, as sea level rises the barrier island becomes increasingly vulnerable to high-energy storm waves that breach the dune line and permit water and sand to rush into the marsh-

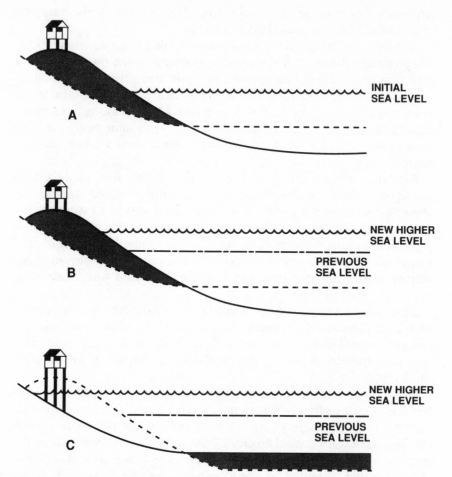

Figure 3.1 *The Bruun rule:* (**A**) *initial condition;* (**B**) *immediate inundation when sea level rises;* (**C**) *subsequent erosion due to sea level rise. A rise in sea level immediately results in shoreline retreat due to inundation, shown in the first two examples. However, a 1-meter rise in sea level implies that the offshore bottom must also rise 1 meter. Waves will erode the necessary sand from upper part of the beach as shown in part* (**C**).

SOURCE: *Adapted from Barth and Titus (1986).*

land or maritime vegetation behind the dunes. As this process repeats itself, the seas gradually wash over much of the island, causing new dunes to form slightly landward of their old positions. Gradually the island shifts landward. This process is generally called retreat, rather than erosion, because the island retains its mass. In the past, entire islands have rolled over in one century.[3]

There is also a possibility that the barrier island will drown in place. As the sea rises, the barrier could be eroded on both the sea and bay sides until sand supplies are exhausted. Then the sea level finally overtakes the island's growth and the island drowns in place. Regardless of whether the island rolls over or drowns in place, consequences to barrier islands from sea level rise are enormous.

In areas experiencing retreat or erosion, high-intensity storms often flood and tear apart developments located in the dune systems. Moreover, such developments may interfere with the natural replenishment process of the dunes, and seawalls built to protect such developments can actually accelerate erosion. Consequently, Congress in 1982 enacted the Coastal Barrier Resources Act to discourage development on undeveloped barrier beaches. Federal flood insurance, federal grants and loans for housing, as well as federal funding for infrastructure such as bridges and roads, are no longer available for 186 barrier islands and beaches from Maine to Texas.

The California coastline is dominated by beaches and cliffs. Of California's nearly 1,100 miles of coastline, approximately 950 miles (86 percent) are undergoing erosion.[4] Of this, the U.S. Army Corps of Engineers has labeled approximately 125 miles as critical, meaning that shorefront buildings and infrastructure are threatened. Sea level rise, the blockage of sediment sources, the existence of near-shore sinks—all contribute to the erosion along California's coast.[5]

COASTAL CLIFFS

A higher sea level will also affect the erosion rates of cliffed coasts, such as those in New England and on the West Coast. These cliffs vary in composition; they can be extremely resistant or made of loose sedimentary material that can be eroded back many feet during a storm. Rising water will increase the rate of cliff erosion for those cliffs made of loose material.

Wave erosion occurs both from pressure exerted by wave impact and from abrasion by sand and gravel carried in the wave. Wave erosion undercuts cliffs, which may remain unsupported for a period of time before collapsing in large segments. Moreover, ocean spray deposits salt on the cliff face, which can chemically weaken it and thus contribute to erosion.

In some areas, eroded material has collected at the base of cliffs, protecting them from undercutting. A rise in sea level could wash away this material, leading to undercutting of the cliff at the rising waterline. As the cliffs erode, development on and behind them becomes vulnerable both to landslides and to serious storm damage.

ESTUARIES

Should projected changes in precipitation patterns from global warming materialize, the destructive impact of sea level rise upon estuaries could be compounded. Estuaries, areas where fresh water meets and mixes with salt water, are extremely dynamic yet sensitive ecosystems that depend on maintenance of regular tidal patterns and salinity. When the sea level rises or freshwater inputs decrease, the so-called salt wedge—the area of mixing between salt and fresh waters—moves farther upstream, increasing the average salinity in the brackish water system. Moreover, with a reduction in freshwater flow, the estuary is starved of essential terrestrial nutrients. Such deprivation affects the overall balance of life in the system.

Changes in estuary composition can affect the surrounding terrain as well. Saltwater intrusion may appear in groundwater, increasing the cost of its treatment for human consumption. Benthic (bottom) substrate will tend to become anaerobic (lose oxygen), and heavy metals sequestered in the substrate will be liberated, allowing sulfur cycles to become dominant. Particulate and soluble organic matter inputs will be reduced or loosely aggregated and deposited near shore rather than being dispersed. Salt-tolerant mosquito and fly populations may increase. Schistosomiasis (a parasitic infection) might become rampant. Noxious salt-loving organisms might move farther upstream. These organisms include the destructive clam *Teredo* (shipworm), the small crustacean *Limnoria* (gribble), and barnacles, all of which destroy wooden docks and boats; dangerous stingrays, which damage clam beds; oyster drills, which prey upon oysters; and sharks, which would pose a potential threat to human water sports.[6]

Estuaries are traditionally seen as effective systems for the dispersal of pollutants. Well-mixed estuaries with rapid flushing rates can dilute and spread out pollutants quickly. Reduced circulation from increased salinity should, in theory, diminish the ability of estuaries to perform this function. The pollutants that would then accumulate could pose a significant threat to humans living in the many urban centers (Baltimore, New York, San Francisco) that border on estuaries.

The alteration of estuaries will severely affect many commercially important fisheries. Some 66 to 90 percent of U.S. fish stocks depend on estuaries for some part of their life cycle, particularly during spawning and the early life stages that are sensitive to water temperature, pH, salinity, and food supply.[7] Sea level rise and alterations in freshwater runoff caused by climate change could affect all four of these factors.

A large number of commercially valuable fish and shellfish species rely on estuaries as nursery grounds for their larval and juvenile stages. The synergistic effects of sea level rise and altered freshwater runoff could be devastating to the delicate larvae and juveniles of many species of fish and shellfish. It is therefore important to maintain freshwater flow into estuaries at constant levels. We should discourage any further alterations to river flow—including increased water consumption, damming, and channelization—or we will risk the loss of billions of dollars in revenue from both commercial and sport fisheries.

OTHER COASTAL WETLANDS, MARSHES, AND MANGROVES

Coastal wetlands are of ecological importance for a variety of reasons: They are regions of high productivity, nurseries for commercially significant fish, and rookeries for waterfowl. Coastal wetlands act as filters for pollutants such as heavy metals, radioisotopes, fertilizer runoff, and sewage effluent. They also act as sponges for floodwaters, thus protecting coastlines from storms and high tides. Efforts to place a dollar value on wetlands estimate that they provide an annual return equivalent to $5,500 per acre.[8] The importance of wetlands protection has been recognized at both the state and the national levels. Both have implemented specific wetland protection programs.

Many of the wetlands along the nation's coasts have been eliminated or severely restricted due to filling or dredging for harbor and port development. Many of the remaining coastal wetlands are degraded due to urban runoff, sedimentation, and reduced freshwater inflows. In California, since the turn of the century, harbor and port development has destroyed about 50 percent of the state's coastal wetlands and about

they used to, and there are fears that shellfishing will soon be a thing of the past.[12] The current threat to the Louisiana bayous is saltwater intrusion into the wetlands. This intrusion is caused by a number of localized factors in addition to global sea level rise. Oil companies have built canals to allow their barges access to the petroleum riches of the bayous. These canals cause increased land subsidence and reduced silt replacement. Furthermore, fresh water from the Mississippi used to flood the marshes seasonally, pushing out the salt water. But now the river rarely floods because of levees constructed to protect riverside development.

A different variation on the same theme threatens the shorelines of the Florida Keys and the Everglades, the only habitats in North America of three tree species known as mangroves. The mangroves, which bind the shorelines with densely intertwined root systems, rely for survival on periodic flushing of water. They are extremely susceptible to alterations in the water balance—and particularly vulnerable to constant inundation. Mangroves contribute to the coast in many ways. The trees protect the shoreline from erosion due to wave action. They also act as natural recycling plants, forming an important part of a delicately balanced nutrient system for the marine life in Florida's coastal waters. Although mangroves have demonstrated some ability to migrate landward in times of gradual sea level rise, they can only do so if the areas behind them are undeveloped.

CORAL REEFS

Warming-induced sea level rise may also cause increased wave action and erosion on reefed coasts such as Florida and Hawaii. Reefs reduce the intensity of incoming waves and thus protect the shore from erosion and storms in the same way that barrier islands protect mainland coasts. A rapidly rising sea would probably outpace the ability of the reef to grow upward, allowing waves to wash over the reefs and penetrate the beach, thus increasing erosion.

A warming of the atmosphere will probably cause a warming of the upper layer of the ocean as well. An ocean warming of only 1° C can kill corals and other reef organisms. Scientists in Miami and Panama have conducted experiments demonstrating the sensitivity of corals to temperature changes, especially increases in warmth. Professor Peter Glynn of the University of Miami Rosenstiel School of Marine and Atmospheric Science says that "scientists have believed that the warm waters of El Niño [off Peru] cause the death of Pacific corals and biological

disturbances. But now by comparing our laboratory results with our field studies affected by the 1983 El Niño, we are able to prove that assumption is correct." Moreover, says Glynn, "while cool currents are thought to be responsible for limiting coral reef development, severe sea warming may also limit reef growth and has to be considered an important factor in the extinctions of reef-building organisms."[13]

Corals have significant ecological and commercial value. They provide the basis for the thriving sport-diving and recreational fishing industries in many southern states, while also serving as collection sites for marine tropical fish species that support the multimillion-dollar aquarium business. Moreover, corals may serve as nurseries for many ecologically and commercially important fish species. The destruction of corals would disrupt the food chain and result in the death of an entire ecosystem, accompanied by the loss of millions of dollars.

Direct Effects on Human Settlements

DAMAGE TO INFRASTRUCTURE

The term infrastructure collectively refers to the nation's highways, bridges, waterways, ports, airports, mass transit systems, facilities for water supply and hazardous waste storage, and their associated maintenance systems. This aging infrastructure will be placed under tremendous strain by a rising sea level of 0.7 meter coupled with the other effects of climate change. A policy of deferred maintenance necessitated by years of budget tightening has caused serious deterioration in the system. A 1988 study published by the National Council on Public Works Improvement found that America's infrastructure is inadequate to sustain future economic growth.[14]

Currently most of the funding for public works comes from local governments, which spent a total of $49.2 billion nationwide on operation, maintenance, and capital improvements in 1984. The federal government followed with $25.9 billion, and states contributed $2[?] billion. While local governments have enjoyed a great deal of autonomy with their public works in the past, in the future they may need more monetary and technical assistance to deal with the effects of a 0.7-m global sea level rise and climate change upon their crumbling infrastructures. A renewed federal effort must be made to list priorities

provide guidance to local governments. The potential costs to the nation's infrastructure from a 0.7-meter sea level rise by the year 2050 are staggering.

Water Management

Salt water is the most common and serious pollutant of fresh groundwater in coastal aquifers. Many examples already exist of rapid saltwater contamination of water table aquifers by human activities. Large-scale construction of canals in southern Florida caused salt water to penetrate into previously freshwater areas, an effect somewhat analogous to sea level rise. Over several years, dense salt water gradually replaced fresh groundwater below the canals. Within fifty years, salt water had intruded up to 13 kilometers (8 miles) in the shallow aquifer near Miami.[15] Similarly, on Long Island, New York, the freshwater/saltwater interface advances 3 to 60 meters per year, depending on local pumping conditions.[16]

At present, a major objective of Florida water management is to prevent further intrusion of salt water into the permeable aquifers. Scientists project that 60 percent of the Everglades National Park would suffer saltwater intrusion with a 1-meter sea level rise.[17]

Sewer Systems

Sewer systems provide for drainage of surface water from streets in the event of a rainstorm. Currently, many city sewer systems rely on gravity drainage—water flows downhill from the streets into the sewers, then continues downhill toward some outfall area. Should the sea level rise and saturate the groundwater to any extent, it could limit the effectiveness of gravity drainage systems and necessitate the installation of mechanical pumping stations to aid drainage of water.

Sewer structures with long operating lives should not be built in erodible sediment adjacent to recreational beaches. The San Francisco Westside Sewer Transport in California is a 2.5-meter concrete culvert designed to provide wet-weather sewer capacity for the city of San Francisco. Buried 400 feet inland along Ocean Beach, a heavily used recreation area, the transport was designed to last 100 years at a cost of $100 million. Yet the rate of erosion caused by current sea level trends would damage the transport by the year 2025.[18] The cost of beach renourishment to protect the structure would range from $11 million to $74 million depending on the rate of erosion. Plans must now be

developed to protect the structure from the effects of a 0.7-meter sea level rise and from increasingly frequent and destructive storms.

Waste Disposal and Storage

When comparing flood elevations between different areas, design engineers set standard recurrence intervals. These are phrased in terms of 1, 10, 25, 50, or 100-year flood elevations. A 10-year flood elevation has a 10 percent chance of occurring in any one year, while a 100-year level has a 1 percent chance of occurring in any year. The 100-year level, commonly used as a safe standard for designs, has been adopted by the National Oceanic and Atmospheric Association, the Army Corps of Engineers, and the Federal Insurance Administration.

As of 1982, there were 1,100 active hazardous waste disposal sites nationwide located within the 100-year floodplain. The EPA does not limit location of such facilities on the floodplain—provided that they are designed, constructed, operated, and maintained to prevent washout of any hazardous waste by a 100-year flood or that the owner or operator demonstrates that, in the event of a flood, the waste would be removed to a safe area before floodwaters reached the facility.

A 0.7-meter rise in sea level would bring more sites into the 100-year floodplain—sites that might not be designed according to Resource Conservation and Recovery Act specifications. Sites already situated within the 100-year floodplain would be exposed to increased risks of inundation and storm damage, shoreline retreat, and changes in the water table. For landfills, the dangers include inundation, waste migration, physical erosion, and saltwater intrusion. For toxic storage facilities, tanks might overflow, spill, or float away if not secured properly. Moreover, floating debris, saltwater corrosion, or increased hydrostatic pressure could cause structural damage. Municipal waste incinerators could get waste in the storage and operating components of the facilities. Not only could structural damage result, but increased salt could corrode components of the system. Considering the wide variety of toxic and carcinogenic chemicals that are routinely stored in these facilities, these threats should not be taken lightly.

Highways, Roads, and Bridges

A 0.7-meter global sea level rise would inundate, weaken, and erode coastal roads. Low-lying roads would be especially jeopardized during storms, risking the lives of motorists. Bridges would be threatened as well. A 0.7-meter global sea level rise would increase bridge structural

load, as well as water-scour bridge foundations. The deterioration of the
nation's roads and bridges has been amply demonstrated in recent years
by the collapse of several major spans—notably the Mianus River
Bridge in Connecticut and the Schiharie Creek Bridge on the New York
State Thruway. Capital spending on roads and bridges by the federal
government, which had declined since a peak in 1968, was finally
increased in 1985. It was an increase in absolute, not relative, terms,
however. With respect to the number of vehicle miles traveled, spending
is still declining. In 1987, a federal Highway Administration survey
found that nearly 250,000 of the nation's highway bridges (about 42
percent) were "deficient," with faults limiting their load, serviceability,
or safety. The cost of repairing these defects might be as much as $51.4
billion. Sea level rise will accelerate the rate of deterioration and thus
require increased funds.

HUMAN SUFFERING AND LOSS OF LIFE

Each year floods threaten the lives of millions of people throughout the
world. Indeed, they have been responsible for some of the worst natural
disasters in history. Despite lessons of the past, people continue to settle
and build upon floodplains. The Dutch live behind an elaborate system of
massive dikes and levees in the face of the persistent threat of a massive
storm flood. The first such flood in the Netherlands, in 1285, killed
50,000 people. Since then, comparable floods have been occurring
roughly every ten years. The largest flood in recent years happened in
1953; it took 1,835 lives and caused $1 billion (1953 dollars) in damage.
    The United States exhibits a similar lack of respect for floodwaters.
Since 1925, more than 4,000 people have died in floods in the United
States and the threat to human lives is still increasing. Estimates of annual
flood losses were $1 billion in the 1960s (1967 dollars) but had risen to $5
billion by 1985. Despite the constant risk of flood disaster, many people
still choose to develop and live on land within the floodplains.

Engineering Responses and Implications

Not only would efforts to stabilize the nation's entire coastline be prohibitively expensive, but the use of hard measures, such as seawalls,

could actually accelerate the loss of valuable ecological and recreational resources. Priority should therefore be given to land use adaptation and to ecologically sound measures, avoiding nonsustainable rigid solutions where possible. In some cases, however, coastal cities will simply have to be protected by structures.

Planners will have to choose among three basic options: (1) stabilizing the shoreline by erecting walls (hard stabilization); (2) raising the land and nourishing the beach with added sand (soft stabilization); or (3) allowing the shoreline to retreat and adapting to it. The appropriateness of each option depends on a number of factors, including the nature of the coastal ecosystem, the size and nature of the human settlement, the area's coastal hazard vulnerability, the relative importance of its resources, the short-term and long-term costs and benefits of the option, the equity of the option in terms of taxpayer dollars, and the physical attributes of the shoreline system.

In most cases protecting an undeveloped shoreline with hard structures such as seawalls would be inappropriate because this measure is neither ecologically sound nor affordable. Although areas that rely on beaches for tourist income might choose to rebuild them, this strategy is enormously expensive, requires yearly replenishment, and may not last past the next severe winter storm. Major cities, however, will probably need to build engineering structures to protect their populations from the severe effects of a 0.7-meter sea level rise and climate change.

HARD STABILIZATION

Hard stabilization entails the construction of groins, bulkheads, seawalls, and revetments. The performance of each type of coastal structure is affected to varying degrees by sea level rise, which causes overtopping and submergence. Groins are wall-like structures extending into the water perpendicular to the shore; they trap sand that is being naturally transported along the shore in the littoral drift. Bulkheads and seawalls, most often taking the form of a vertical wall facing the sea, are placed on shorelines above the mean high-water line to protect upland structures. A revetment is a system of loose or interlocking units laid on a slope below the mean sea level; it protects loose shoreline from the direct scour of the ocean waves.

In recent years, these hard stabilization measures have become the focus of a major debate between coastal engineers and geologists. Coastal geologists claim that the construction of permanent hard structures adjacent to or within the softer coastline almost always accelerates erosion. Coastal engineers reply that these protection measures need not accelerate erosion if they are correctly built. Coastal engineers agree, however, that if the structure is poorly designed or affects the littoral sand transport in some way, there will indeed be accelerated erosion. And coastal geologists concede that because seawalls are almost always constructed in areas that already have an erosion problem, it is sometimes difficult to separate prewall from postwall erosion. In the face of a sea level rise of 0.7 meter, there will *always* be postwall erosion.

Despite the obvious major engineering considerations, decisions concerning hard stabilization measures will be partly determined by economics. The long-term costs to urban cities of maintaining the structures must be considered when analyzing the costs and benefits. Moreover, such factors as design life, responsibility for inspection and maintenance, and environmental effects will increase the cost.

SOFT STABILIZATION

In the case of beach nourishment, sand is pumped from an offshore location through huge pipes to the shore. When no care is taken to ensure proper placement and types of sand, the multimillion-dollar investment can be washed away in a matter of months, as happened in Ocean City, Maryland. Moreover, if there are no efforts to reduce the turbidity caused by the project, the local marine environment may be severely harmed. Concerns for the local fauna—especially coral formations—inhibit many beach nourishment schemes in Florida.

A final consideration involves the thorny issue of "sand rights." Which communities own which sand supplies? With a finite amount of offshore sand available for such projects, who decides which communities should be allowed access to the sand and which should be excluded? These issues are beginning to turn up in the courts and may become much more frequent in the years to come.

RETREAT AND RELOCATION

The only strategy of adaptation to sea level rise that is universally endorsed by planners and scientists is that of retreat and relocation. Certainly, retreat from the shoreline is always the least damaging option for the environment. Moreover, it is economically desirable—once the retreat has been accomplished, no more investment is needed for maintenance or additional flood protection measures. In communities that do not wish to inflict environmental damage, a planned and orderly retreat from the shore is the most suitable response to rising sea levels.

Retreat can be either a gradual, planned process or a response to a catastrophe. In the latter case, of course, the retreat is a direct response to a natural occurrence. A planned retreat, on the other hand, might take several years to implement and would involve the creation of new zoning and setback laws. These laws would first establish mandatory setback lines to prohibit new development seaward of areas subject to the effects of a 0.7-meter sea level rise as well as increasingly frequent and destructive storms. They might be followed by government-sponsored relocation incentives. Insurance coverage and mortgages will gradually become unavailable for structures located in these areas, as business learns to appreciate the increased risks.

Unwisely, most people will be tempted to rebuild houses when they are destroyed by storms or erosion. They do not understand, for example, that if a structure was destroyed by a "100-year storm" in 1985 and then rebuilt in 1986, it is not guaranteed safe until 2085. In fact, the storm is just as likely to occur in 1989 as it is in 2085. There is a 1 percent chance, by definition, of a 100-year storm occurring in any given year.

Chapter 4

FEDERAL POLICY

So far, the United States' policy on controlling or limiting global warming has been extremely cautious. Consequently, the development of U.S. policies for adapting to the *effects* of global warming—such as sea level rise and increased storminess—have proceeded very slowly as well. With the exceptions of the U.S. commitment to 1987 Montreal Protocol reductions of chlorofluorocarbons and to tree planting, United States policy on global warming has been limited to the encouragement of scientific research and the analysis of policy alternatives.

Presidential and Congressional Action

The Bush administration's policies on global change—or lack thereof—have been controversial. Although some cabinet officials—such as EPA Administrator William Reilly and Secretary of State James Baker—appear to support action to curb greenhouse gases, other powerful administration officials—such as White House Chief of Staff John Sununu—are reluctant to commit the United States to greenhouse gas reductions because of fear of economic dislocations.

In November 1989, the Bush administration joined with over seventy countries attending the Ministerial Conference on Atmospheric Pollution and Climate Change in Noordwijk, the Netherlands, in calling for a stabilization of carbon dioxide emissions as soon as possible and for the investigation of quantitative targets to limit or reduce such emissions. However, the United States led efforts to block specific commitments to specific dates and reduction amounts. The EPA's William Reilly, as well as Dr. Allan Bromley, science and technology adviser to President Bush, noted at the conference that the United States, which at that time was spending $500 million per year on the study of issues related to climate change, planned to increase its spending to about $1 billion in fiscal year 1991. Moreover, the White House restated its commitment to a total phaseout of chlorofluorocarbons by the year 2000.

In February 1990, however, President Bush appeared to retreat from the White House commitment to stabilize carbon dioxide emissions as soon as possible. In his remarks at the opening session of the Intergovernmental Panel on Climate Change (IPCC) meeting in February 1990, Bush indicated that his administration was not prepared to announce new policy initiatives on global warming. At the same time, his administration submitted to Congress a Department of Energy budget request reflecting a 46 percent reduction in spending on energy efficiency in fiscal year 1991.

The Bush administration's paralysis with respect to the causes of global warming has had a chilling effect on the development of policies to respond to the *effects* of global warming. By failing to signal that the threat of global warming is serious enough to require actions to limit it, this country's top leaders are sending a message to the federal agencies as a whole that they need not plan for land use adaptation to coastal threats from global warming and sea level rise.

Many major bills to control global warming have been introduced in Congress, but none has been enacted into law. Proposed amendments to the Coastal Zone Management Act to deal specifically with sea level rise also have been introduced in Congress. This legislation would help coastal states adapt to the adverse effects of sea level use and climate change—primarily by encouraging state coastal zone management programs, as part of their federally required significant improvement tasks, to study, to develop within five years, and to implement management plans addressing the adverse effects on the coastal zone from sea level rise and global warming. The plans would include the implementation of policies (1) to consider sea level rise from global warming in the siting of new infrastructure investments and new large-scale developments which have long-life expectancies, such as sewage treatment plants, industrial plants, and hazardous waste facilities, (2) to create buffer zones for those wetlands that could migrate landward as sea level rises, (3) to ensure that structural protection measures are ecologically sound and avoid nonsustainable rigid solutions, and (4) to require building setbacks and standards that will minimize the adverse effects of sea level rise and storms on human settlements and ecosystems.

The Council on Environmental Quality

The Council on Environmental Quality (CEQ) is the implementing agency for the National Environmental Policy Act of 1969 (NEPA) and operates under the executive branch. NEPA and its implementing regulations require federal agencies to evaluate the reasonably foreseeable environmental impacts of every federal agency proposal for legislation and other federal actions significantly affecting the quality of the human environment.[1]

Although past sea level rise and future sea level rise from global warming have "reasonably foreseeable environmental impacts" for many proposed federal actions along the coast, no federal agency has expressly discussed sea level rise from global warming in any environmental impact statement (EIS). Indeed, the U.S. Army Corps of Engineers has not required sea level rise from global warming and other climate change effects to be evaluated in designing major coastal projects.[2]

Similarly, no federal agency has evaluated global warming or ozone depletion in an EIS since 1978, although an informal survey by the Council on Environmental Quality identified three such statements prepared in the 1970s that discussed the impacts of climate change due to the greenhouse effect and ozone depletion.[3] These statements include the Federal Aviation Administration's EIS on the applications of Air France and British Airways to operate the Concorde in the United States (1975); the Food and Drug Administration's EIS on its regulatory action to limit use of chlorofluorocarbons as propellants in aerosols (1977); and the National Aeronautic and Space Administration's EIS on the Space Shuttle Program (1978).

In 1988, the CEQ held a series of informal hearings on the ozone depletion issue with a view to issuing a guidance letter on the need for agencies to evaluate global ozone depletion and climate change in their EISs.[4] On November 9, 1988, the Natural Resources Defense Council urged the CEQ to require all federal agencies to examine fully the effects of climate disruption, including accelerated sea level rise, for all major federal actions significantly affecting the quality of the human environment.[5] More than a year later there had been no response. One of the Reagan administration's last actions was to direct the CEQ not to issue the guidance letter but to hold it for the Bush administration. The CEQ under the Bush administration has reportedly circulated a proposed guidance to the federal agencies but has encountered delay at the Justice Department and the Department of Energy. To date, there still has been no final action by the CEQ.

Federal Agencies

There are five federal agencies primarily responsible for the coastal environment: the Environmental Protection Agency; the Department of Commerce (which administers the Coastal Zone Management Act); the Federal Emergency Management Agency (which administers the federal Flood Insurance Program); the U.S. Army Corps of Engineers; and the Department of Interior (which manages national parks and wetlands). All of these agencies operate under statutory authorizations that arguably require them to take sea level rise from global warming into account in their activities. Only the Environmental Protection Agency has done so to any significant degree, however, and even its activities have been limited to scientific and policy research.

THE ENVIRONMENTAL PROTECTION AGENCY

Congress directed the Environmental Protection Agency to be the lead agency on global change and charged EPA with implementing the 1987 Global Climate Protection Act.[6] The act sets goals for U.S. policy—slowing the rate of increase of concentrations of greenhouse gases in the atmosphere in the short run and stabilizing or reducing atmospheric concentrations of greenhouse gases over the long run. In December 1989, the EPA submitted a report to Congress on the effects of global climate change on the forests, lakes, wetlands, agriculture, human health, and other environmental issues.

EPA has operated a Sea Level Rise Project under its Office of Policy Planning and Evaluation since the early 1980s. The principal activity of this office has been to prepare several important policy and research papers. These studies have pointed out the policy choices the nation must make in the face of rising seas. The EPA's policy research has examined the effects of sea level rise on various coastal regions, focusing especially on flooding and shore erosion. In a 1983 study, the EPA projected high, medium, and low estimates for future sea level rise and concluded that the East Coast is slowly sinking.[7] Recently it has applied these findings to Ocean City, Maryland, a city that has had substantial beach erosion during the past century. The report's authors conclude that the relative sea level rise at Ocean City will be 6 to 8 inches greater than the global sea rise per century.[8]

Similarly, the EPA has investigated the physical effects of various sea level rise scenarios on the shoreline at Sea Bright, New Jersey. This study predicts that the yearly probability of intermediate to serious damage to the area's protective seawall will increase significantly if no shore protection measures are implemented. The study also predicts significantly increased flooding as the sea level rises. The study reaches the very questionable conclusion, however, that "the most reasonable solution" for protecting Sea Bright at its present location would be to raise the elevation of necessary roadways, vital structures, and utilities. Much of the private housing and commercial buildings might also have to be removed.[9]

The EPA has also studied the consequences of future sea level rise for our nation's wetlands and marshes. Charleston, South Carolina, for example, could lose between 50 and 90 percent of its marshes in the next century. An estimated 2.9-foot sea level rise could result in a loss of 50 percent of Charleston's marsh by the year 2075. The area of high marsh would decline from 2,300 to 700 acres, while the area of low marsh would decline from 5,400 to 3,200 acres. With a 5.2-foot sea level rise (the EPA's own high scenario), Charleston could lose 80 percent of its marsh by 2075, with high marsh declining from 2,300 to 700 acres and low marsh from 5,400 to 900 acres.[10]

Similarly, the U.N. Environment Program and EPA conducted a joint study projecting dramatic wetlands loss nationwide. Applying the Sea Level Affecting Marshes Model (SLAMM), the study concludes that the Florida Keys will rapidly become open water after 2075 and that the southern Everglades will also disappear, although wetlands may increase where inland areas are inundated. With few exceptions, the Gulf Coast marshes will gradually disappear until the barrier islands are breached; marshes not protected by dikes will then decline precipitously. The Mississippi Delta will also experience large-scale loss of marshes at an accelerated rate early in the next century. Losses in New England and the West Coast will be less dramatic; present New England salt marshes will disappear with no compensating gain in new marsh area after 2075, while West Coast marshes should persist throughout most of the next century.[11]

Moreover, the EPA commissioned three separate studies to assess the regional implications of sea level rise. These studies included an engineering analysis examining the cost of holding back the sea versus the cost of not holding back the sea, a wetland model examining the amount of coastal wetlands that would be lost with and without shore protection, and an examination of the cost of protecting the open coast and the possibility of barrier island breakup. As well, the wetlands modelers looked at the additional wetland loss that would result from such a barrier breakup.[12]

These research efforts are clearly a major step in reaching solutions for the problems posed by global warming and sea level rise. Unfortunately, the EPA has thus far failed to translate these findings into action. For example, its wetlands division has yet to plan a vigorous wetland protection strategy.

THE DEPARTMENT OF COMMERCE

The Department of Commerce is responsible for administering the Coastal Zone Management Act (CZMA). Enacted by Congress in 1972 and most recently reauthorized in 1986, the CZMA is the only piece of national legislation that attempts to deal comprehensively with coastal problems. Under this act, federal funding encourages coastal states to develop and implement coastal zone management programs. To date, twenty-nine of the thirty-five eligible coastal states and territories have developed such programs.

While most coastal state management programs do not currently take into account the environmental effects of sea level rise from global warming, the CZMA does provide a useful framework for encouraging coastal states to address these impacts. Section 306(a)(3) of the CZMA requires that each coastal state spend an increasing proportion of its federal grant money on activities that will significantly improve the coastal zone management objectives: "the management of coastal development to minimize the loss of life and property caused by improper development in flood-prone, storm surge, geological hazard, and erosion-prone areas and in areas of subsidence and saltwater intrusion, and by the destruction of natural protective features such as beaches, dunes, wetlands, and barrier islands." If the state fails to make significant improvements in achieving these objectives, the secretary of commerce is required to reduce the financial assistance provided to the state by up to 30 percent.[13]

As defined in the regulations of the CZMA, a "significant improvement" includes an accomplishment that addresses any of the objectives of Section 303(2)(A)–(I) by substantially expanding the scope of the state's coastal zone management program.[14] Consequently, state adoption of new enforceable policies that take into account sea level rise from global warming, the development of site-specific management plans for vulnerable areas, the development of natural resource information related to sea level rise, and the development and implementation of strategies to address sea level rise could all be considered significant improvements under the regulations.[15] The state is responsible for initiating improvements,[16] but the Office of Ocean and Coastal Resource Management (OCRM), which administers the CZMA, can also exercise leadership by encouraging states to incorporate such tasks in their financial assistance applications. So far, a number of coastal states have done so: California, Delaware, Florida, Hawaii, Louisiana, Maine, Maryland, Massachusetts, New Hampshire, New Jersey, New York, North Carolina, Oregon, Rhode Island, South Carolina, and Washington.[17] Current proposals to amend CZMA to encourage greater state planning for sea level rise and other climate change effects would assist in coastal state land use adaptation.

THE FEDERAL EMERGENCY MANAGEMENT AGENCY (FEMA)

Currently, the primary means for flood management is state participation in the National Flood Insurance Program pursuant to the National Flood Insurance Act of 1968.[18] Under the National Flood Insurance Program, the federal government provides subsidized flood insurance in communities that adopt adequate land use controls and enforcement measures for regulating development in flood hazard areas. Typically, local regulations require developers to build in a fashion that would minimize damage in a 100-year flood. (The 100-year flood, recall, is defined as a flood for which there is a 1 percent chance of occurring in any given year.) Although several of these state flood management programs include criteria that will be useful in planning for sea level rise, state flood planning is largely based on FEMA's 100-year flood insurance rate maps—which are based only on historical flood data.[19] Until these maps incorporate projections of sea level rise, local flood management planning will be ineffective in mitigating the flood-related effects of sea level rise.[20]

Another important oversight in the National Flood Insurance Program has been its failure to consider coastal erosion when setting premiums. However, the 1987 two-year reauthorization of the program does allow for advance insurance payments for relocating or demolishing structures in imminent danger of destruction from erosion or flooding. Congress also provided for further flood insurance coverage on reconstruction and relocation landward of the predicted thirty-year erosion line.

Relocation is a sensible response to increasing coastal erosion, flooding, and storm damage. Certainly homeowners who move out of hazardous coastal areas reduce the risk to their lives and property. Relocation also benefits the environment. Shoreline development often involves the destruction of existing dunes, which provide important habitat for birds and other wildlife and may shelter adjacent wetlands. Moving intensive developments away from the coast can also reduce coastal water degradation, which harms marine life, by leaving a buffer to filter land-based sources of pollution and runoff.

No assessment of the effects of sea level rise and climate change on the program has been undertaken. By FEMA's own admission, it has not performed particularly well with respect to managing losses. An interagency task force under FEMA's guidance concluded that the current annual flood losses were "unacceptable." The task force further recognized the need for "better decisions affecting the use of our nation's floodplains, reduced losses of life, property, and natural values, and reduced burden upon government to compensate for losses caused by the unwise decisions of individual citizens as well as governments."[21] This need is especially acute in view of the effects of the probable 0.7-meter global sea rise by the year 2050.

Although FEMA in the past achieved some reduction in flood losses—and succeeded in securing local government participation in managing the flood-prone areas as well as cost sharing by local governments and floodplain owners—its current policies are wholly inadequate for the scope of the danger to coastal settlements presented by an accelerated global rise in seas. Protecting present development from flooding and preserving the natural values of urban floodplains (flood storage, aquifer recharge, water quality, wildlife habitat) will be an enormous challenge in the coming century.

Present-day coastal development is largely exempt from any regulations concerning flood protection. A 1983 survey of local and state governments concluded that while 60 to 100 percent of new development was subjected to regulations concerning flood protection in floodplain areas, only 40 to 90 percent of existing development was forced to comply. State and local governments have been hesitant to ask real estate owners to retrofit their properties with flood mitigation measures. As a consequence, only 21 percent of local officials rated present flood hazard management as very effective in "reducing the exposure of existing development to flooding."[22]

Faced with the effects of global warming, however, not only the case for floodplain control but also the case for floodplain *redefinition* becomes compelling. FEMA has not conducted strategic assessments of how the program could be managed to minimize damage from shoreline retreat caused by both present and future rates of sea level rise, largely because the National Flood Insurance Act does not require strategic assessments of long-term issues.

THE ARMY CORPS OF ENGINEERS

The U.S. Army Corps of Engineers continues to advise its engineers to design projects based exclusively on past sea level rise rather than take into account sea rise from global warming. A Corps guidance letter states that "hydraulic designers should not make projections of the degree of either global or relative sea level rise on other than a historic basis." Corps of Engineers policy is that "until substantial evidence indicates otherwise, the Corps will maintain the procedure of considering only local regional history of sea level change to project a rise or fall for a specific project."[23]

THE DEPARTMENT OF INTERIOR

The U.S. Department of Interior is charged with management of the nation's public lands, including the national parks, the national seashores, the outer continental shelf lands, and lands managed by the Bureau of Land Management. Global warming will seriously stress the national parks when changes in precipitation cause drought and ecosystem migration. It will also cause erosion, flooding, and inundation of the national seashores and may dry up the nation's grazing lands. As well, it may increase opposition to oil and gas leasing of outer continental shelf lands, since oil and gas burning contribute to global warming. The Interior Department has initiated an internal analysis of the impacts of global warming on public lands.

Chapter 5

CHALLENGES FACING THE COASTAL STATES

States are beginning to recognize that erosion, loss of wetlands, saltwater intrusion into groundwater supplies, flooding, storm damage, and other problems are related to sea level rise.[1] To avoid or mitigate the effects of accelerated sea level rise and climate change, the need for state planning and direct action is pressing. Nearly 102.5 million people now live within 50 miles of U.S. coastlines.

Despite the magnitude of potential problems, the challenges presented by a sea level rise are not unmanageable. In fact, sixteen coastal states have already initiated studies of sea level rise and its effects on their coastal regions.[2] Moreover, although sea level rise was not taken into account in designing most coastal zone management programs, a number of laws and policies already in place could be used to address the problems that must be faced. Three jurisdictions have adopted new policies responding to sea level rise: Maine, South Carolina, and San Francisco Bay.[3] Tables indicating coastal states' responses can be found in Appendixes D and E of this book.

State and Local Studies

Several state and local studies highlight the dramatic effects of sea level rise on coastal areas. In parts of the country as diverse as Hawaii and Delaware, California and Massachusetts, local studies have consistently proved that these effects will be devastating throughout the coastal United States. Although each area has its unique characteristics and responses, there are several common themes.

First, the shape of the coastline will be altered, resulting in the loss of coastal land and marshes, disruption of transportation, and eventually the loss of upland areas. Second, changes in tidal circulation will cause increased wave action, impeded drainage, and increased flooding. Finally, the chemical and mineral makeup of the water will change. In many areas, the rate of sedimentation will slow down. Increased salts and chlorides will harm irrigated crops and poison groundwater, thus destroying the drinking water supply for local populations. The following studies illustrate these drastic alterations in various locations throughout the country.

THE CALIFORNIA COAST

Presently, over 60 percent of California's population lives within the coastal counties, and this percentage is expected to rise in the future.[4] The California coastline stretches for 1,100 miles and includes cliffs, beaches, and a variety of wetland types. Approximately 86 percent of this coastline is presently eroding. Almost 10 percent of the eroding areas already have hardened shoreline protective structures, and additional sections employ nonstructural protective methods.

Because California has a geologically active coastline, local land subsidence and uplift must be taken into account in any consideration of relative sea level rise. Areas of Santa Barbara and Ventura counties are undergoing uplift, for example, while areas of Humboldt and San Mateo counties are undergoing subsidence. Even areas undergoing rapid rates of coastal uplift, however, will experience beach erosion, bluff retreat, and submergence of lowland areas. The greatest impact will occur in low-lying areas, which would be completely inundated. Moderately severe impacts will occur in areas with broad beaches or fragile coastal bluffs protected by loose rocky material. The least impact will occur in areas of resistant, steep coastal cliffs.

Assuming that a low sea level rise by the year 2050 would be 0.1 meter, a moderate rise by 2050 would be 0.3 meter, and a high rise by 2050 would be 0.45 meter, the study finds that:

• Sea level rise will exacerbate existing erosion problems and cause erosion and cliff retreat in currently stable areas.

- Coastal beaches will be reduced in size. Coastal beach retreat between 30 to 200 feet can be expected by the year 2050.
- In some areas, 35 to 100 percent of existing wetlands could be lost if development prevents upland migration.
- Higher water levels and wave forces will increase the uplifting forces on piers and erosion of foundation supports.
- Harbors may experience greater wave action, and higher water levels will adversely affect loading and unloading of cargo ships.
- Existing protective structures such as breakwaters, seawalls, and revetments will not be able to provide their current level of protection due to increased forces, erosion, and overtopping.

SAN FRANCISCO BAY

This study projects a sea level rise of about 1.2 meters over the next century and concludes that the physical character and shoreline of San Francisco Bay and the Sacramento River/San Joaquin River Delta would be dramatically altered.[5] Assuming that it would only be feasible to maintain levees to protect urban development, the study concludes that levees which protect salt ponds and delta islands would eventually fail: "The area of the Bay-Delta system then would approximately double over the next century, with the most dramatic change being the creation of an inland sea in the Delta." The study predicts numerous other effects as well, including:

- A change in tidal circulation
- An increase in salinity in Suisun Bay and the Sacramento River Delta that would "drastically affect California's irrigated agriculture in the San Joaquin Valley and southern California's water supply"
- Possible reduction of sedimentation rates resulting from more sediment sinks in newly inundated salt ponds, agricultural lands, and diked wetlands
- Increased capture of sediment by the delta "sea" and submergence of shallow subtidal areas that currently act as seasonal storage reservoirs
- Increased wave action
- Impeded drainage
- Increased flooding of adjacent low-lying areas
- Possible contamination of groundwater
- Reduced areas of salt marsh, brackish marsh, and seasonal wetlands

THE DELAWARE ESTUARY

This study assumes a 0.7-meter rise in sea level in the Delaware estuary by the year 2050 and a 2.5-meter rise by 2100.[6] The study concludes that sea level rise and climate changes caused by the greenhouse effect are likely to have profound impacts on the quality and quantity of water in the Delaware River Basin. Specifically, the study finds that sea level rise could substantially increase the salinity of the Delaware estuary in the next century. Without countermeasures, a recurrence of the 1960s drought combined with sea level rise could produce the following scenarios:

- A 0.7-meter rise would move the salt front upstream to river mile 100 (currently at river mile 93) and increase sodium concentrations to more than 50 ppm (the New Jersey drinking water standard) during 15 percent of the tidal cycles at Philadelphia's Torresdale intake.
- A 2.5-meter rise would move the salt front upstream to river mile 117, increase salinity to 1,560 ppm at the DRBC salinity control point (the current level is 136 ppm),[7] and increase sodium concentrations to more than 50 ppm during 50 percent of the tidal cycles at Philadelphia's Torresdale intake.

Accelerated sea level rise could also threaten the New Jersey aquifers recharged by the Delaware River by increasing river water concentrations of chloride during a drought. Again, a repeat of the 1960s drought would produce the following conditions:

- A 0.7-meter rise would increase chloride concentrations to 350 ppm in the water recharging the Potomac–Raritan–Magothy aquifer. (The EPA standard for drinking water is 250 ppm.)
- A 2.5-meter rise in the worst month of drought would increase chloride concentrations to greater than 1,000 ppm in 75 percent of the recharge and greater than 250 ppm in 89 percent of the recharge.

The study recommends identifying potential reservoir sites and preserving them from development to ensure that they are available to increase reservoir capacity. It also recommends initiation of a regional study to examine the potential impacts of precipitation as well as sea level rise for the Delaware estuary and adjacent river basins.

HAWAII'S COASTAL ZONE

This study focuses on the effects on Honolulu of several scenarios of sea level rise over the next 100 years:[8]

- A 0.6-meter sea level rise would result in the following effects: major flooding threats during storm conditions; loss of Waikiki Beach; increased losses of property to storm waves; increased cost of shoreline protection; and temporary disruptions of transportation at Honolulu Harbor, international airport, and three major surface thoroughfares.
- A 1.5-meter rise would lead to the following results: coastal hazards threatening not only shorefront areas but also properties up to a mile inland; at a minimum, the enormous if not prohibitive cost of stabilizing the entire Honolulu waterfront; loss of valuable urban land; geographic isolation of Waikiki; and prolonged disruption of surface, air, and maritime transportation systems. Moreover, much of the present urban district may become uninhabitable.

Sea level rise would also affect Hawaii's groundwater by increasing leakage to the sea and causing greater saltwater infiltration in wells. The overall result would be a smaller supply of usable groundwater.

MASSACHUSETTS' COASTAL ZONE

This study concludes that by the year 2025 there will be startling amounts of upland loss in Massachusetts.[9] Even assuming that only the historical rate of sea level rise continues, the following amounts of land would be lost:

Rate of Relative Sea Level Rise	Amount of Upland Lost Between 1980 and 2025
Present rate	3,000 acres
0.3-meter rise[a]	7,500 acres
0.5-meter rise[a]	over 10,000 acres

[a] Includes constant rate of local coastal subsidence.

The report assigns oceanfront property a nominal value of $1 million per acre; at that price the economic loss for a sea level rise of 0.5 meter by 2025 would be $10 billion.

These five studies point out the magnitude of the problems the states will face as a result of rising sea levels, as well as the urgent need to develop state policies dealing with the threat.

State Environmental Quality Review

One way to evaluate the effects of climate change on the states' coastlines is through consideration of sea level rise in the environmental impact statements required under state statutes. Analysis of the issue in an EIS ensures that state agencies considering the propriety of a particular development are aware of the implications that climate change may have on the proposed development, as well as the development's impact on the environment. Unlike other statutes, which may require amendment in order to address the matter of sea level rise, statutes requiring an EIS can and should currently require consideration of sea level rise for activities in the coastal zone.

For example, New York's State Environmental Quality Review Act (SEQRA) requires all state and local agencies to prepare an EIS on any action they propose or approve that may have a significant effect on the environment.[10] There are certain criteria for determining environmental effects: a substantial increase in potential for erosion, flooding, or drainage problems; the creation of a hazard to human health or safety; effects on important freshwater wetlands; effects on recharge of underground aquifers; effects on water pollution control; and a substantial change in the capacity of land or other natural resources to support existing uses. Sea level rise and its consequences are clearly relevant to each of these criteria. Moreover, New York's SEQRA is substantive: Its clear expression of legislative intent is that state and municipal agencies give "due consideration ... to preventing environmental damage."[11]

A number of other states have similar statutes.[12] Because of North Carolina's Environmental Policy Act of 1971, for example, a state agency must include an EIS for any recommendation for legislation or actions that will involve expenditure of public money for projects and programs "significantly affecting the quality of the environment" of North Carolina. The EIS must include detailed statements of the environmental impact of the proposed action; any significant, unavoidable, adverse environmental effects; proposed mitigation measures; alternatives to the proposed action; the relationship between the short-term uses of the environment involved in the proposed action and the maintenance and enhancement of long-term productivity; and any irreversible and irretrievable environmental changes.[13] In addition, the act authorizes local governments to require any special-purpose unit of government or private developer of a major project to submit an EIS.[14]

Permits for Development

A number of states require permits for development within the coastal zone and limit development to certain uses. Although existing permit requirements and development limitations were not designed to mitigate losses from sea level rise, they reflect general policies aimed at averting harm and loss of property as well as protection of environmental resources. Thus they offer a mechanism that states could use for minimizing the costs of sea level rise.

Delaware's Beach Preservation Act states:

Beaches of the Atlantic Ocean and Delaware Bay shoreline of Delaware are hereby declared to be valuable natural features which furnish recreational opportunity and provide storm protection for persons and property, as well as being an important economic resource for the people of the State. Beach erosion and shoreline migration occur due to the influence of waves, currents, tides, storms, and *rising sea level*. These natural forces have created, and will continue to alter, the beaches of the State. Development and habitation of beaches must be done with due consideration given to the natural forces impacting upon them and the dynamic nature of those natural features. The purposes of this chapter are to enhance, preserve, and protect the public and private beaches of the State [and] to mitigate beach erosion.[15]

The regulations implementing the act provide that the geology, geomorphology, meteorology, and hydraulics of the area must be taken into account in considering permit applications.[16]

The policies enunciated in other state coastal programs and implementing regulations provide potential means for considering sea level rise and its effects with respect to coastal development. For example, the guidelines of Louisiana's Coastal Resources Program state:

> It is the policy of the coastal resources program to avoid the following adverse impacts. To this end, all uses and activities shall be planned, sited, designed, constructed, operated, and maintained to avoid to the maximum extent practicable significant . . . land loss, erosion, and subsidence[;] increases in the potential for flood, hurricane, or other storm damage; or increases in the likelihood that damage will occur from such hazards.[17]

Erosion Mitigation

Apart from general development restrictions, many states have restrictions aimed at specific environmental hazards. Among the primary effects of sea level rise will be the accelerated erosion of shorelines, beaches, and already threatened barrier island systems. The methods chosen for controlling or adjusting to this erosion will themselves affect the profile of the coasts. Because structural control measures often accelerate erosion and block sediment deposition, and because their effectiveness is generally short-term, nonstructural approaches to dealing with erosion are preferred. Of course, nonstructural measures such as relocation and setback requirements may not always be feasible, particularly in highly developed areas.

STRUCTURAL REQUIREMENTS

A few states prohibit the use of structural erosion control measures. North Carolina, for example, prohibits the construction of any structure designed to stabilize or harden the oceanfront shoreline. Such structures include wooden bulkheads, seawalls, rock or rubble revetments, jetties, groins, breakwaters, concrete-filled sandbags, and tire structures.[18] Instead North Carolina favors beach nourishment and relocation of structures (such as buildings) as erosion response measures.[19] Similarly, no new seawalls are permitted under Maine's Coastal Sand Dune Rules. (A seawall is defined as an embankment, vertical wall, or other barrier built for the purpose of preventing shoreline erosion.)[20] Excerpts from Maine's Coastal Sand Dune Rules can be found in Appendix F of this book.

A number of states do, however, allow the use of structural erosion control measures, generally under limited circumstances. (See Appendix F.) New York's Coastal Erosion Hazard Areas Law may be particularly useful as a model for considering the appropriateness of such measures in light of sea level rise. The law requires that any "activities, development, or other actions in such erosion hazard areas should be undertaken in such manner as to minimize damage to property and to prevent the exacerbation of erosion hazards." Further, activities in an erosion hazard area "may be restricted or prohibited if necessary to protect natural protective features or reduce erosion impacts."[21] Publicly financed structures will be erected only where necessary to protect human life, existing investment in development, or new development that must be located in the erosion hazard area or adjacent coastal waters in order to function. Both publicly and privately financed erosion protection structures must minimize damage to other man-made property or to natural protective features or other natural resources. Officials must carefully weigh the long-term costs of such structures against public benefits before construction is undertaken.[22]

As currently interpreted, New York's Coastal Erosion Hazard Areas Law requires consideration of future erosion in determining whether a structure will have a reasonable probability of controlling long-term erosion for at least thirty years; it also requires that a long-term maintenance program be provided for the structure.[23] Thus the Hazard Areas Law might be interpreted to require consideration of sea level rise in determining whether a structure will have a reasonable probability of controlling long-term erosion and in determining a maintenance program. Other states' structural control methods are outlined in Appendix E.

SETBACK REQUIREMENTS

An important tool in preserving beaches and minimizing the costs of sea level rise will be setback requirements—laws restricting construction or development to areas beyond reasonably anticipated points of erosion. Such requirements will not only help to preserve beach areas as sea level rise hastens erosion. They will also lessen the economic costs of sea level rise by minimizing loss of structures or the need for huge investments to protect them. Most coastal states have adopted some type of setback requirement. (See Appendix E.) While the majority did not adopt setback requirements with explicit reference to sea level rise, some state laws do require setbacks to take into account anticipated erosion.

Rules under Maine's Sand Dune Law actually require that sea level rise be taken into account in determining building sites.[24] The Board of Environmental Protection must "evaluate proposed developments with consideration given to future sea level rise."[25] The law does not permit new developments if, "within 100 years, the project may reasonably be expected to be damaged as a result of changes in the shoreline."[26] Moreover, the rules allow construction within the sand dune system of buildings greater than 35 feet in height above existing grade or covering a ground area greater than 2,500 square feet (about the size of a single-family home) only if the applicant for a permit can demonstrate by clear and convincing evidence that the site will remain stable after allowing for a 0.9-meter rise in sea level over 100 years.[27]

Another example is Florida's Beach and Shore Preservation Act.[28] Although it does not currently address sea level rise, the act holds potential for incorporating its effects in coastal planning. The act authorizes the Florida Department of Natural Resources (DNR) to establish construction control lines that define the portion of the beach-dune system that is subject to severe fluctuations based on a 100-year storm surge, storm waves, or other predictable weather conditions. DNR may establish a construction control line further landward than the impact zone of a 100-year storm surge, provided the line does not extend beyond the landward toe of the coastal barrier dune structure that intercepts the 100-year storm surge.[29] Control lines are subject to review at the discretion of DNR if hydrographic and topographic data indicate that shoreline changes have rendered established control lines ineffective.[30] The construction control lines are not setbacks. Rather, they define the areas that require permits and special structural design considerations to ensure protection of the beach-dune system, any proposed structures, and adjacent properties.[31]

In addition to the construction control lines, which define the areas of extreme impacts, 1985 amendments to the Beach and Shore Preservation Act prohibit all construction seaward of the construction control lines except for shore protection structures, minor structures, piers, and certain single-family homes.[32] Homeowners must demonstrate that their homes will remain seaward of the seasonal high-water line for thirty years after the date of a permit application.[33] The Department of Natural Resources makes thirty-year erosion projections on a site-by-site basis.[34] These projections, however, are based on historical measurements of horizontal shoreline change. Thus they do not account for the projected additional sea level rise from global warming and the consequent erosion, which are crucial elements in making such limits effective. Furthermore, a construction prohibition based on thirty-year erosion projections is unlikely to protect structures that are intended to remain standing for more than thirty years. Since sea level rise is expected to begin accelerating around the year 2020, such structures could be seriously damaged.[35]

In 1988, largely in response to sea level rise studies and research, South Carolina established a setback baseline which would be recalculated every five to ten years, based upon the average erosion rate for the previous forty years.[36] South Carolina's process will thus gradually factor in accelerating sea level rise. As time passes, the provision of the new law that prohibits rebuilding destroyed properties seaward of the setback line has become quite controversial—especially in light of the extensive damages Hurricane Hugo inflicted on South Carolinians shortly after the new act went into effect. The law specifically provides for court determination of whether a rebuilding prohibition constitutes a taking without compensation. If a taking is found to have occurred, the state can either issue a building permit or provide reasonable compensation for lost land use.[37]

Flood Protection

As sea level rises, states will not only face eroding coastlines but also the effects of increased flooding caused by storm surges, heightened storm intensity, and the loss of storm protection formerly provided by the eroding beaches. Currently, the primary means for flood management is state participation in the National Flood Insurance Program administered by FEMA.[38] Although most state flood planning is based largely on FEMA's historical flood data, a few states do incorporate sea level rise into their planning.[39] Of particular interest are regulations recently adopted pursuant to Maine's Sand Dune Law.[40] These regulations provide that "in order . . . to prevent creation of flood hazards, the Board [of Environmental Protection] will evaluate proposed developments with consideration given to future sea level rise."[41]

The rules govern development within the sand dune system based on flood insurance rate map zones. The V-Zone is the area within the 100-year flood that is subject to additional hazard from high-velocity water due to wave action.[42] No new structures or additions to existing structures are permitted in the V-Zone.[43] The A-Zone is the area of the 100-year flood.[44] The B-Zone covers those areas between A-Zones and V-Zones and the limits of the 500-year flood.[45] New buildings and additions to buildings in the A-Zone and B-Zone must be constructed so that the lowest portion of the structural members of the lowest floor is at least 1 foot above the elevation of the 100-year flood and must be adequately constructed to withstand a 100-year storm.[46] All multiple-unit buildings consisting of three or more dwelling units must be constructed so that elevation of the first inhabited floor is at least 4 feet above the 100-year flood elevation.[47] As with the erosion regulations, "no building greater than thirty-five feet in height above existing grade, or covering a ground area greater than 2,500 square feet (approximately the size of a single-family home), shall be constructed in a sand dune system unless the applicant demonstrates by clear and convincing evidence that the site will remain stable after allowing for a three-foot rise in sea level over one hundred years."[48]

Another useful example of state flood planning occurs in New York. Its Floodplain Management Regulations govern development in flood hazard areas by permit, using 100-year flood data.[49] The regulations establish minimum elevation levels for new structures according to base flood elevation,[50] the level of the 100-year flood,[51] or mean high tide,[52] depending on the type of 100-year flood data available.[53] Any new major developments or subdivisions within a flood hazard area must be consistent with the need to minimize flood damage.[54] All utilities and service facilities in new developments must be located and constructed to minimize or eliminate flood damage.[55] A new or replacement water supply system or sanitary sewage system must minimize or eliminate the infiltration of floodwater into that system and must minimize or eliminate discharges from the system into floodwater.[56] The regulations also prohibit any new facilities, petrochemicals, or hazardous or toxic substances in any flood hazard area.[57] Although these regulations do not currently cover sea level rise, they could easily be adapted to do so.

Other state policies that encourage preventive actions might also be useful in planning for sea level rise. Massachusetts, for example, has developed several restrictions on development in flood-prone areas. The state's Coastal Zone Management Program (CZMP) policies include:

- Restricting new development in V-Zones, barrier beaches, sandy beaches, primary dunes, and certain salt-marsh areas to primarily nonstructural uses
- Controlling new development in contiguous upland areas (extending 100 feet inland of the limit of the 100-year flood) to ensure that existing hazards are not exacerbated and that the proposed uses or activities are appropriate in light of the risks of damage
- Ensuring that proposed development in intertidal areas or offshore in coastal water bodies will not exacerbate existing erosion or flooding hazards in adjacent or downcoast areas[58]
- Acquiring undeveloped hazard-prone areas for conservation or recreation use

Furthermore, the Massachusetts State Building Code requires new construction or substantial improvements within a coastal high-hazard area to be certified by a registered professional engineer or architect. Such structures must be elevated so that the lowest portion of the structural members of the lowest floor is at the 100-year flood line; they must also be securely anchored to withstand high-velocity water and hurricane wave wash.[59] Finally, Gubernatorial Executive Order No. 181 prohibits the use of state and federal funds for construction projects that encourage growth and development on barrier beaches. The order assigns the highest priority for use of disaster assistance funds to relocate willing sellers from storm-damaged barrier beaches and directs that state acquisition of barrier beaches be a priority.

Protecting Natural Resources

Although destruction of property may be the most immediately obvious result of sea level rise, the destruction of natural resources will be equally devastating. In several states, current regulations could be used or adapted to protect our wetland, groundwater, and surface water supplies.

WETLANDS PROTECTION

As sea level rises, the most important means of protecting wetlands will be to ensure that they can migrate landward. Thus the most effective wetlands regulations provide for a limited development buffer zone adjacent to wetlands.

California's Coastal Act provides that development in areas adjacent to environmentally sensitive habitats such as wetlands, parks, and recreation areas be sited and designed to prevent impacts that would significantly degrade such areas.[60] (The California Coastal Commission's statewide interpretive guidelines consider "adjacent areas" as those at a distance of up to 500 feet from a wetland.)[61] Although this provision of the Coastal Act is currently interpreted to mean degradation as a direct result of a development,[62] the statutory language requiring that development "be compatible with the *continuance* of such habitat areas" could provide a framework for consideration of future wetlands migration.

Maryland's Chesapeake Bay Critical Area Protection Program establishes a minimum 100-foot buffer from the edge of tidal waters, tidal wetlands, and tributary streams to the head of tide.[63] New development activities—including structures, parking areas and other impervious surfaces, mining and related facilities, and septic systems—are prohibited in the buffer zone except for those necessarily associated with water-dependent facilities.[64] Local jurisdictions must expand the buffer beyond 100 feet to include contiguous sensitive areas—such as steep slopes, hydric soils, or highly erodible soils—whose development may have an impact on streams, wetlands, or other aquatic environments. In the case of contiguous slopes of 15 percent or greater, the buffer must be expanded 4 feet for every 1 percent of slope, or to the top of the slope, depending on which is greater.[65]

New York's Tidal Wetlands Act and regulations categorize the state's wetlands into zones. The different zones—coastal fresh marsh, intertidal marsh, coastal shoals, bars and flats, littoral zone, high marsh or salt meadow, and adjacent areas—are regulated more or less stringently, depending on their environmental and storm protection value. A table of "compatible" and "incompatible" uses regulates development in the different zones.[66] "Adjacent areas" are defined as any land either immediately adjacent to a tidal wetland, within 300 feet of the most landward boundary of a tidal wetland, at the seaward edge of the closest existing substantial man-made structure, or at the elevation contour of 10 feet above mean sea level.[67] These lands are afforded the lowest level of protection under the statute.[68] Notably, however, the function of the adjacent areas is "to serve as buffers to protect the character, quality, and values of tidal wetlands that adjoin or lie near these areas."[69] Although not intended to provide a buffer for migration, this language might be interpreted to cover migration necessary to protect the wetlands from sea level rise. Legislative clarification would strengthen the provision's usefulness, however.

Finally, New Jersey's Rules on Coastal Resources and Development require a buffer that could protect future wetlands migration. The rules provide that "all land within 300 feet of wetlands . . . and within the drainage area of those wetlands comprises an area within which the need for a Wetlands Buffer shall be determined."[70] Development is prohibited in a wetlands buffer unless it can be demonstrated that the proposed development will cause only a minimum feasible adverse impact on the wetlands and on the natural ecotone (an ecological community of mixed vegetation formed by the overlapping of adjoining communities) between the wetlands and the surrounding upland.[71] The precise geographic extent of the buffer on a specific site is determined on a case-by-case basis and could be determined with reference to probable future sea level rise from global warming.[72]

WATER QUALITY AND SALTWATER INTRUSION

As sea levels rise, intrusion and pollution will threaten the quality of drinking water, as well as water used for agriculture and industry. Several issues will have to be addressed: the effect of pumping on water tables, location of new drinking water pumping stations, placement of sewage and solid and hazardous waste treatment facilities, and reservation of currently available lands for future reservoirs should saltwater intrusion render existing water supplies impotable. Several state water quality control policies provide useful models for addressing some of the water quality problems that can be expected as a result of sea level rise.

California's County Water District Law,[73] for example, creates water districts with broad powers to manage current water supplies and ensure future supplies. A water district "may do any act necessary to furnish sufficient water in the district for any present or future beneficial use."[74] Among the actions a district may take are storing water for the benefit of a district, conserving water for future use, and appropriating, acquiring, and conserving water and water rights for any useful purpose.[75]

Florida's Water Resources Act of 1972[76] requires the Department of Environmental Regulation to study existing water resources; means and methods of conserving and augmenting such waters; existing and contemplated needs and uses of water for the protection and procreation of fish and wildlife; irrigation; domestic, municipal, and industrial uses; and all other related subjects, including selection of reservoir sites.[77] Further, the department must work with the governor to formulate an integrated plan based on such studies for the use and development of the state's waters.[78] The Department of Environmental Regulation must also identify those areas of the state where saltwater intrusion is a threat to freshwater resources and report its findings to water management districts created by the Water Resources Act.[79] At the request of a water management district (or on its own initiative when it is determined that saltwater intrusion has become a matter of emergency proportions), the department must establish a "saltwater barrier line." Inland of this line, no canal may be constructed or enlarged and no natural stream may be deepened or enlarged without a dam or other barrier structure at or seaward of the saltwater barrier line.[80]

Florida's Water Resources Act also creates a Water Management Lands Trust Fund.[81] The trust fund is for the use of water management districts to acquire lands necessary for water management, water supply, and the conservation and protection of water resources, as well as for management, maintenance, and capital improvements of lands included within water management plans.[82] Similarly, Georgia's Ground Water Use Act of 1972 requires the Board of Natural Resources to regulate groundwaters "to protect against or abate salt-water encroachment."[83] The act also prohibits any person from withdrawing, obtaining, or utilizing groundwater in excess of 100,000 gallons per day for any purpose other than agriculture without a permit.[84]

HAZARDOUS WASTE FACILITIES

Because states are only beginning to appreciate that sea level rise and its consequences are accelerating, they have paid little attention to the effect this rise and increased storms will have on hazardous waste facility siting. Consideration of sea level rise in future siting of hazardous waste facilities will be critical to ensure that such facilities are not exposed to flooding or wave attack, saltwater intrusion, or submergence and to prevent leaching of wastes into groundwater. Careless placement of treatment facilities will have tragic effects on our natural resources.

New Jersey's Major Hazardous Waste Facilities Siting Act is useful because it provides a mechanism for introducing consideration of sea level rise into the siting process.[85] The act instructs the Department of Environmental Protection to prepare criteria for the siting of new major hazardous waste facilities and prohibits location or operation of such a facility within certain designated flood hazard areas, wetlands, and any area where the seasonal high-water table rises to within 1 foot of the surface.[86] Moreover, the criteria must prevent any major adverse environmental impact, including major degradation of the surface or ground waters of New Jersey.[87]

Other state hazardous waste management statutes and regulations might also be used to limit siting of hazardous waste facilities in areas subject to adverse effects of sea level rise. Connecticut, for example, requires its Hazardous Waste Management Service[88] to prepare a management plan, an inventory of preferred areas for hazardous waste management facilities, and a report on the suitability of each preferred area if private waste facilities are not meeting state needs.[89] In evaluating and selecting the candidate sites the service must consider the risk to the local public health, safety, and welfare, including the risks from water, air, and land pollution and adverse effects on agriculture and natural resources.[90]

Finally, Maine requires that all hazardous waste facilities be located to assure protection of human health and welfare and the environment. Such protection includes preventing adverse effects on ground and surface water quality, especially those due to the migration of waste constituents in the subsurface environment.[91]

Two Case Studies

Two areas present unique circumstances: Louisiana is a special case because it will be one of the states likely to suffer the greatest losses, due in part to geography and in part to human activities. The Great Lakes are unique because their problem in a warmer world is more likely to be reduced precipitation and a decline in lake levels.

LOUISIANA

While the harsh effects of sea level rise induced by global warming probably will not confront most coastal states for several years, the devastating consequences of relative sea level rise are already apparent in Louisiana. The rate of land loss within the Louisiana coastal zone is approximately 55 square miles per year (nearly 100 acres per day).[92] Moreover, Louisiana's barrier islands are eroding in places at rates up to 20 meters (65.6 feet) per year.[93] Although these difficulties are primarily due to localized factors, they present a striking picture of what a global sea level rise portends for coastal communities. Other coastal states would do well to examine the causes of sea level rise and land loss in Louisiana, as well as the state's proposed measures for mitigation.

In many ways, the problems Louisiana is now facing are the results of unique human activities rather than environmental events. This is particularly true of wetland areas. For example, the harnessing of the Mississippi River for flood control and navigation purposes has deprived the delta plain of delta-building sediments.[94] Withdrawal of oil and gas from the delta has increased subsidence of the land surfaces, contributing to an increase in relative sea level rise.[95] Canals constructed for oil and gas industry activities and for navigation have caused direct wetland loss due to displacement by water and soil deposits. Moreover, these canals have contributed to indirect marsh loss as a result of alterations in hydrology, loss of over-marsh sheet flow, increased saltwater intrusion, impoundment of marshes, and acceleration of local subsidence.[96] Finally, outright reclamation of wetlands (by dredging, filling, or draining) to create land for residential, industrial, and agricultural purposes has directly reduced the wetlands areas of Louisiana.[97]

Another factor in the loss of wetlands has been an increased rate of shoreline erosion. The increase is due primarily to a natural process—an increased frequency of hurricane landfall—but it is exacerbated by human activities such as jetty construction, reef removal, and sand mining, as well as the indirect impact of human activities such as reduction in sediment and accelerated subsidence.[98] The rate of coastal erosion is also speeded by the disintegration of Louisiana's storm-protective barrier islands and beaches. These barrier shorelines, formed when the Mississippi River naturally abandoned delta lobes,[99] are particularly sensitive to sea level rise and storm surge. Long-term relative sea level rise, land loss, and erosional shore-face retreat will eventually separate the barrier shoreline from the mainland and form a barrier island arc.[100] Subsequently, storm impacts will overcome the arc's ability to maintain its integrity near the earth's surface. Submergence eventually will occur and an inner-shelf shoal will form.[101]

The effects of these various processes on Louisiana's wetlands and barrier islands have been nothing short of devastating. Louisiana has lost over 1 million acres of coastal wetlands in this century and at the present rate of acceleration could lose another million acres in the next twenty to twenty-five years.[102] The barrier islands declined in area by 37 percent between 1890 and 1979—from 98 to 52 square kilometers (38 to 20 square miles).[103] The life expectancy for the Isles Dernieres barrier system is only fifty years.[104] All of these problems will, of course, be exacerbated by a global sea level rise, which is expected to result in a relative Louisiana sea level rise of 1.42 to 3.53 meters (4 feet 8 inches to 11 feet 7 inches) in the next century.[105]

Is the situation hopeless? While land loss in Louisiana cannot be totally arrested, a number of ameliorating measures have been proposed or have already been undertaken. A number of these measures are specific to the situation in Louisiana, but many of them will be useful for protecting other coastal states. In 1981, Louisiana established a Coastal Environment Protection Trust Fund and appropriated $35 million for projects to combat coastal land loss.[106] In 1985, the state approved a Coastal Protection Master Plan, which sets out a strategy for dealing with land loss. Projects developed under the Master Plan will control erosion, restore and nourish barrier islands, deter saltwater intrusion, and restore wetlands.[107]

Several projects are designed to restore barrier islands and shorelines. A project to stabilize the Isles Dernieres barrier island chain (located in Terrebonne Parish) involves the restoration of low dunes and washover beaches.[108] For a width of up to 1,000 feet, engineers will place dredged material along 16 miles of barrier shoreline. The stabilized dunes and enlarged island base will reduce the chain's susceptibility to storm breaching.[109] Similar methods will be used to restore and stabilize Shell Island (Plaquemines Parish).

A similar project has already stabilized the Fourchon Island shoreline (Lafourche Parish). This 1985 project closed old Pass Fourchon, relocated a beach road, and restored dunes. Moreover, restoration crews combined their work with damage repair from the hurricane season. A hydraulic dredge pumped approximately 700,000 cubic yards of beach material into a soil retention area; the material restored the beach and dune. To further stabilize the dune, the Port Commission plans to revegetate it and provide toe protection for the seaward retaining structure.[110]

Louisiana has also planned marsh restoration and management projects. In St. Bernard Parish, Trust Fund money will be used to repair or construct levees and water control structures. This construction will aid the implementation of wetland management plans to offset habitat degradation for two environmental management areas.[111] Within the Pointe au Chien Wildlife Management Area (Terrebonne Parish), the

Montegut Marsh Restoration will use levees, fixed crest wiers, and a flapgate culvert to stabilize water levels and reduce salinity and turbidity. The project's planners intend to reduce present rates of marsh loss, bring about increased production of desirable plant species, and increase fish and wildlife habitat and production.[112] A similar project in the environmentally sensitive La Branche Wetlands, located on the fringe of Lake Pontchartrain (St. Charles Parish), will combat severe erosion and shoreline breaching. The project will use shoreline restoration and stabilization or shore protection measures. If successful, it will protect both the shoreline and the 3,000 acres of deteriorating wetlands that lie behind it.[113]

As these measures illustrate, we *can* take steps to mitigate some of the harmful effects of sea level rise. But Louisiana's situation also demonstrates that the consequences of sea level rise in many places will be dramatic. The costs of reducing these effects are high, and substantial advance planning will be the most helpful measure of all.

THE GREAT LAKES

The Great Lakes present a contrasting case study on the effects of global warming. While most areas will experience an accelerated rise in sea level, current scientific studies project a *decrease* in Great Lakes water levels.[114] Several climatic effects are expected to contribute to this lowering of lake levels. Higher temperatures from global warming will likely cause increased evapotranspiration from the land surface, thereby lessening runoff into lakes.[115] Likewise, higher temperatures will result in increased evapotranspiration from lake surfaces.[116] Moreover, likely changes in precipitation patterns and possible increases in consumptive use could further reduce water supplies to the Great Lakes system.[117]

Estimates of the amount of change in Great Lakes levels vary with estimates of global warming. Under one scenario, which assumes a 4.3° to 4.8° C carbon-dioxide-induced warming over the next 100 years, the net basin supply of the Great Lakes (defined as the sum of the precipitation on the lakes and the runoff into the lakes, minus the evaporation from the lake surfaces) will decrease 20.8 percent.[118] Under a more moderate scenario, which assumes a 3.1° to 3.7° C warming over the next 100 years, the decline in net basin supply would be 18.4 percent.[119]

There are possible benefits of global warming in the Great Lakes region—for instance, a longer recreational season and diminished ice coverage—but scientists also predict numerous adverse effects of lower lake levels. Decreased channel depths in the Great Lakes/St. Lawrence Seaway would necessitate extensive dredging in both the connecting channels and the major harbors.[120] Much of the dredged material is highly contaminated and would create a disposal problem. Reduced flows and a lower water surface would also contribute to lower hydro-power production, possibly necessitating the use of less clean and more costly alternative sources of power.[121] Although higher temperatures would deteriorate water quality, global warming could increase consumption of water—and lake levels would drop even further.[122] Other projected effects include drying out of closed marshes, migration of open marshes to new lake levels, and reduced agricultural yields.[123]

Lower lake levels will also hinder navigation. The lakes will be unable to accommodate larger vessels, and deep-draft vessels will lose carrying capacity. These problems will increase the number of trips necessary, the transportation costs, the potential for vessel damage, and the operation and maintenance costs for many vessels.[124] Areas accessible to small craft, such as recreational vessels, would also diminish.[125]

Thus while the effects of global warming on the Great Lakes water levels are projected to be markedly different from those on sea level, the Great Lakes region will likely experience numerous coastal changes. And, as in areas threatened by sea level rise, the areas surrounding the Great Lakes and the users of the lakes should begin planning for adjustment to these expected changes.

Chapter 6

INTERNATIONAL
EFFECTS AND POLICIES

Since 1988, scientists, ecologists, economists, and policymakers from around the world have met in Toronto, New Delhi, Tokyo, Noordwijk, Cairo, and Miami to analyze the effects of climate change on coastal settlements and ecosystems. Each time they have concluded that climate change poses severe risks to human populations located along coastlines and to the ecosystems on which they depend.[1] The 1989 Cairo Compact, which focuses on the effects of global warming and sea level rise on coastal areas, fisheries, and marine resources, cautions that a 1-meter sea level rise might involve the displacement of 300 million people. For the Statement and Resolutions of Working Group 3 of the Cairo Conference, relating to coastal areas, fisheries, and marine resources, see Appendix A of this book.

Building a Consensus

In 1988, the World Meteorological Organization (WMO) and the United Nations Environment Program (UNEP) established the Intergovernmental Panel on Climate Change (IPCC) to develop an internationally accepted strategy for addressing climate change and its effects. The IPCC reported its findings on August 31, 1990, at a meeting hosted by Sweden. The IPCC's Coastal Zone Management Subgroup (CZMS) identified the adaptive options and policy implications of sea level rise and climate change on the coasts. Its goal is to provide information and recommendations to national and international policy centers regarding coastal zone management strategies for the next ten to twenty years, as well as long-term policies dealing with adaptation to the impacts of global climate change, including sea level rise.

The IPCC Coastal Zone Management Subgroup concluded that "it is urgent for coastal nations to begin the process of adapting to sea level rise not because there is an impending catastrophe, but because *there are opportunities to avoid adverse impacts by acting now*—opportunities that may be lost if the process is delayed."[2] The IPCC Subgroup's four national coastal planning recommendations follow:

1. By the year 2000, coastal nations should implement comprehensive coastal zone management plans;
2. Coastal areas at risk should be identified;
3. Nations should ensure that coastal development not increase vulnerability to sea level rise; and
4. Emergency preparedness and coastal zone response mechanisms need to be reviewed and strengthened.[3]

The situations faced by island nations and by many developing nations are especially severe. Island nations cannot always simply retreat inland, and many coastal developing countries have enormous populations living at the very edge of the sea, barely surviving on its resources. These nations do not have the means to feed their populations today, much less to relocate, feed, and house them inland. Furthermore, climate change is expected to increase the exposure of fisheries and other marine resources to solar ultraviolet radiation. It may affect ocean circulation patterns, as well, and lead to increased turbidity, water temperature, and storminess. Moreover, it may lead to the loss of spawning and nursery areas, especially mangroves and seagrass flats, further jeopardizing these nations' sources of food.

Both developed and developing nations need to negotiate a climate convention to slow the warming and limit its effects as soon as possible. Such an international treaty is especially vital to the enormously vulnerable human settlements along the coasts. At the same time, the nations of the globe should be moving to cooperate and assist coastal nations in adapting to sea level rise and climate change. An international coastal protocol should be adopted as part of the convention on climate change. Such a protocol would formally recognize that the adverse effects of climate change may be causing difficulties for coastal nations already and that these effects can only be aggravated in the future. Moreover, such a protocol may persuade recalcitrant coastal nations to become signatories to the convention on the climate. A draft protocol can be found in Appendix B of this book.

Many countries that have no plans to combat sea level rise will experience severe effects from even a moderate rise. Two such countries are Bangladesh and Egypt. Holland, on the other hand, a nation that has met the challenge of rising sea levels for the past thousand years and succeeded in keeping its lowlands dry, may serve as a model for small Western countries.

Three Case Studies

A good way to begin to grasp the magnitude of the likely effects of climate change on coastal communities is to focus on coastal communities which will be particularly hard-hit: Egypt, Bangladesh, and Holland.

THE NILE DELTA

Of Egypt's million square kilometers of land, only the area lining the banks of the Nile River (35,000 square kilometers or 3.5 percent) is densely populated.[4] Nearly 1,400 people cram into every square kilometer of Egypt's breadbasket, relying mainly on farming to provide for their needs. On either end of the Nile's massive delta, the large port cities of Alexandria (in the west) and Port Said (in the east at the Mediterranean entrance to the Suez Canal) harbor 3 million and 800,000 people, respectively. For thousands of years, this nation's people and economy have been tied to the Nile River in some way. Farmers, seafarers, and merchants have relied on the steady flow of the ancient river to sustain their crops and carry their boat traffic. It is only in the last twenty years, however, that the Egyptians have realized that the Nile also continually builds the land, maintaining the level of much of the coast in times of rising sea levels.

The 1964 closure of the high dam at Aswan, some 900 kilometers up the river, cut off the sediment supply to the delta. The combined effects of local tectonic subsidence and global sea level rise caused a pronounced relative rise in sea level that submerged many coastal areas. While the subsidence and global sea rise had been occurring for the past

7,500 years, a constant supply of sediment carried down the giant river had built up the delta at a comparable rate, effectively nullifying the adverse effects. Construction of the dam suddenly ended this process, and today, more than twenty years later, coastal residents feel the full impact of the action.

Should the current trend of relative sea level rise continue in Egypt, 8 to 10 million people may be displaced as flooding waters submerge their land. With a 1-meter rise in sea level, approximately 15 percent of the country's gross domestic product (GDP) will be affected in some way, and 12 to 15 percent of the nation's arable land will be submerged. Given a 3-meter rise in sea level, approximately 20 percent of the GDP will be affected and 20 percent of the arable land will be submerged. The spectacular erosion already occurring along the deltaic coast will intensify greatly.[5]

Clearly such changes would be catastrophic for Egypt, destroying much of the country's farming capabilities, causing widespread hunger, and sparking mass migration away from the flooded lands. Much depends on the ultimate rate of sea level rise, but we should assume a 0.7-meter global rise by the year 2050. Egypt is largely unprepared for this rise, and one can only speculate about the adverse global repercussions of the coming changes in that country.

BANGLADESH

Bangladesh, surrounded by India on the Bay of Bengal, is one of the most densely populated countries in the world. Within its borders 93 million people occupy 143,000 square kilometers at a density of 650 people per square kilometer. The largest city, Dhaka, houses 4 million people, while the next biggest, Chittagong and Khulna, have populations of 1.5 million and 800,000, respectively. The remaining 86 million people are evenly distributed throughout the country, barely surviving on subsistence agriculture. A 1983 study estimated that 85 percent of the population receives less than the 2,122 calories per day necessary for minimal subsistence.[6] The population is poor, overcrowded, often hungry, and growing—at the rate of 2.5 to 3.0 percent a year.[7]

Nearly 80 percent of the land is made up of the complex Bengal delta system, fed by the Ganges, Brahmaputra, and Meghna rivers. Agricultural production on these lands makes up about 55 percent of the country's GDP, and approximately 85 percent of the people depend on agriculture for their livelihood. The overworked lands are protected by the dense forest and mangrove system that lines many of the waterways. If these forests did not exist, storm surges would likely wipe out large sections of the nation's arable land.

A 1-meter sea level rise will affect approximately 9 percent of Bangladesh's people, destroy about 11 percent of the nation's crops, and thus affect nearly 6 percent of the GDP. A 3-meter rise would be much more devastating, affecting 27 percent of the nation's people, 27 percent of the crops, and about 15 percent of the GDP. These numbers do not include the indirect effects of saltwater intrusion into the nation's fresh groundwater sources; intrusion could extend as far as 480 kilometers inland, thereby prohibiting irrigation of crops in that region.[8]

Compounding the adverse impacts of sea level rise on agriculture, the nation's fishing industry could lose as well. The 1.5 million people who depend on fishing for their livelihood currently provide the nation with 80 percent of the consumed animal protein. Roughly 40 percent of the nation's fishing capacity is centered in areas likely to be inundated by a 3-meter rise. While some fishing may be relocated, the significant fraction that depends on freshwater fishing may be lost entirely.[9]

Storm surges pose another major threat to human life in Bangladesh. Historically vulnerable to major cyclonic activity, the country was hit in November 1970 by a devastating storm that killed upward of 250,000 people. A 1985 storm killed 5,000. The combined effects of increased storm surges from sea level rise, the death of protective mangrove buffer zones due to inundation, and conjectured increases in storm power due to the greenhouse effect might spell further disaster for the low-lying areas.

The nation of Bangladesh is neither financially nor socially prepared to cope with the predicted rise in sea level. Such a rise will increase the number of starving people in the country and further weaken its financial position in the world. The consequences of inaction in this nation will be catastrophic. Both a climate convention and a protocol concerning cooperation and assistance in responding to the effects of sea level rise and global warming are vital to Bangladesh's future.

THE NETHERLANDS

The mention of protecting large urban areas against sea level rise inevitably conjures up images of Holland. Those who oppose advance planning for such a rise point to the experience of the Dutch to prove that walls to keep out the sea can be quickly built anywhere. These critics ignore the fact that the Dutch have fought a battle against the sea for centuries. This "ongoing dialogue" has met with partial success, but only at the cost of thousands of lives and billions of dollars.[10]

The earliest walls and dikes in the Netherlands were probably built in the thirteenth century to reclaim land that had been lost to the sea. These walls apparently had little effect, and land was often lost to the sea as quickly as it was reclaimed. In the eighteenth century, the dikes were connected into an integrated system, augmented first by windmills to pump water from behind the dikes and later by mechanical and electrical pumps.

In 1953, a catastrophic storm flooded much of southern Holland, killing 1,850 people. After that storm, the Dutch recognized the necessity of improving their system, and they invested considerable time, money, and careful engineering to construct bigger and better dikes. Their formula for the height of the new walls incorporated historically observed sea level rise and attempted to predict a water level that would be exceeded only by a 1 in 10,000 chance. A dike built to withstand a water level of 5 meters, for example, must actually be much higher than 5 meters in order to resist tidal oscillations, long waves (seiches), and storm surges. Thus the actual formula might resemble the following:[11]

Storm surge level	5.00
Wave runup	9.90
Seiches and gust bumps	0.35
Sea level rise	0.25
Settlement	0.25
Total height	15.75 m above MSL

The major project produced nearly 400 kilometers of dikes and 200 kilometers of dunes on the Dutch coastline to protect the 8 million people who live behind them. The system's annual maintenance cost is about $35 million (1986 dollars)—a cost that is more than made up for by the fact that nearly one-half of the country would be under water if the dikes did not exist.

The Dutch example is illustrative in many ways. The most obvious lesson is that they have apparently won the battle against the sea, succeeding in keeping their land dry by armoring the coastline and building gigantic dikes. Second, the careful and coordinated effort of 1953 has averted subsequent storm or flooding catastrophes. Third, while the initial construction of the dikes required a major capital investment, their subsequent maintenance is an extremely minor sum compared to the benefits they provide.

The fourth point is a cautionary one: It took a tragic storm, which cost the lives of thousands and resulted in millions of dollars worth of damage, to shake the Dutch into a major concerted effort to improve their system. While there is no U.S. situation exactly comparable to that of the Dutch, the disaster of 1953 serves as a warning to the complacent. Will we in the United States wait for a titanic storm that takes thousands of lives before we begin to act against rising sea levels? Clearly, our situation does not warrant the construction of huge dikes and levees. But it does force us to begin planning appropriate countermeasures while there is still time to implement them. If we begin to plan now, we may be able to shorten our future reaction time and thus avert major disasters.

Chapter 7

FINDINGS AND RECOMMENDATIONS

The consequences of global warming and sea level rise on the world's coastal communities and coastal ecosystems will be dramatic and unfortunate. The costs in human suffering and loss of coastal ecosystems will be enormous. Yet we can reduce global warming's adverse effects if we begin to adapt now. It is up to the nations of the world, the Congress, the president, and federal and state agencies to act on that potential.

Findings

Global mean sea level is an indicator of climatic change. During the last 35,000 years, sea levels have dropped and risen markedly as global temperatures warmed and glaciers melted. Sea level can be thought of as the dipstick of climate change. There are several causes of global mean sea level rise: thermal expansion of the upper layers of the ocean; melting of mountain glaciers; melting of the Greenland glacial cap; and discharge from the Antarctic ice sheet.

During the most recent 100 years, the global mean rise in sea level has been about 15 centimeters. The most likely explanation for this rise is global atmospheric warming. Three independent lines of evidence corroborate that global mean sea level has been rising during the past 100 years: tide gauge records; erosion of 70 percent of the world's sandy coasts and 90 percent of America's sandy beaches; and melting and retreat of mountain glaciers. The correspondence between the two curves of rising global temperatures and rising sea levels during the nineteenth century appears to be more than coincidental.

Relative sea level at any given location is affected by a number of factors in addition to changes in global mean sea level: short-term fluctuations due to meteorological phenomena such as those associated with the El Niño warming event; land subsidence such as that associated with groundwater or oil withdrawals; readjustments of the earth's crust due to removal of ice sheets; and tectonic movements. Local and regional subsidence is responsible for a large portion of the relative sea level rise along most of the U.S. coast. Subsidence along the mid-Atlantic coast has amounted to 1 foot for the past century (about ten times greater than the worldwide average). The Atlantic coast average of 2 to 3 feet of beach erosion per year, as well as the Gulf Coast rates in excess of 5 feet per year, must be added to the global trend. The Pacific coast has been stable on average. (Much of the shore is hardrock.)

Large quantities of carbon dioxide and other greenhouse gases—released into the atmosphere from such activities as burning fossil fuels, leveling forests, and producing synthetic chemicals such as chlorofluorocarbons—are expected to raise the earth's average surface temperature by at least 1.5° to 4.5°C in the next century. Many scientists believe that the greenhouse effect is already affecting our climate.

Dramatic reductions in greenhouse gas emissions would slow the rate and magnitude of the warming—and thus the rate and magnitude of sea level rise. Projected emissions of greenhouse gases need to be cut by over 60 percent in order to return concentrations to present day levels. A 22 percent reduction in CO_2 emissions could be achieved by implementing the following package ot measures: higher federal vehicle fuel efficiency standards, federal actions to improve lighting efficiency, higher federal appliance efficiency standards, federal actions to promote industrial efficiency, state building efficiency standards, actions to promote renewable energy sources, a federal conservation reserve forestry program, improved management of forest lands, urban tree planting, and improved mass transit. The United States can achieve further reductions by banning the production and use of chlorofluorocarbons and reducing emissions of methane and nitrous oxide.

Models based on global warming of 1.5° to 4.5° C have predicted a rise in global mean sea level of 0.3 meter (with an uncertainty factor of 0.4 meter) by the year 2050 and a rise of 2 meters by 2100. These models predict ever-increasing rates of sea level rise over the next century; exponential increases will be realized over a longer time frame. While there appears to be a twenty-year lag in the realization of these high rates, it can be viewed either as a planning horizon or as an unwarranted rationale for complacency.

Ongoing rates of sea level rise already pose serious threats to our coasts—for example, 70 percent of the world's sandy coasts and 90 percent of America's sandy beaches are eroding. According to scientists, the sea levels along the U.S. eastern seaboard will rise perhaps an additional 6 inches over the next fifty to sixty years—based on factors other than global sea level rise from global warming. For policy development purposes, a planning horizon of fifty to sixty years is reasonable and prudent for major government investment.

Nations should be planning for a 0.7-meter global sea level rise along their coasts by the year 2050. To this amount, of course, must be added local changes in sea level due to regional subsidence and other factors. A 0.7-meter global sea level rise would produce major losses of valuable natural resources and ecosystems, such as beaches, dunes, estuaries, barrier islands, and wetlands. It would also result in salinization of coastal drinking water supplies; serious threats to already weakened coastal infrastructure such as roads, seaports, mass transit, wastewater facilities, solid and hazardous waste facilities, and powerplants; and potentially enormous human suffering from loss of life and property damage to settlements during storms.

A 1987 report by the National Research Council concluded that many of the adverse consequences of accelerated sea level rise can be reduced, even avoided, by coastal land and water planning that takes the effects of sea level rise and global warming into account. The NRC recommended immediate planning for an accelerated rise in sea level. Efforts to stabilize the nation's coastline would be prohibitively expensive. Moreover, the use of hard stabilization measures, such as seawalls, could actually accelerate the loss of valuable ecological and recreational resources. Priority should thus be given to land use adaptation and ecologically sound protection measures, avoiding nonsustainable rigid solutions where possible.

Building bulkheads and levees, pumping sand, and raising barrier islands for America's most developed coastal miles would cost between $73 billion and $111 billion (EPA 1988 estimate) for a 1-meter rise. The cost of maintaining the country's recreational beaches through sand replenishment is estimated to be $100 billion. Neither figure includes cost estimates for flood damage repairs, for elevating roads and buildings, for modifying or relocating major infrastructure, for the value of lost land and wetlands, for lost recreational opportunities, and for loss of human life or injuries caused by storms.

Since 1988, distinguished scientists, economists, ecologists, and poli-cymakers have met in Toronto, New Delhi, Tokyo, Noordwijk, and Cairo. They have concluded that coastal nations need to begin now to develop strategies for adapting to the adverse effects of accelerated sea level rise and climate change on their coasts. At the state level, there is widespread recognition of the potential effects of global warming on the coasts and the need to begin adapting coastal land uses. Many of the nation's coastal states have begun to assess the effects of sea level rise on their coastal zone management plans. Yet only two of the coastal states (Maine and South Carolina) and one regional authority (the San Fran-cisco Bay Conservation and Development Commission) have actually adopted land use strategies in response to sea level rise.

The ecological and socioeconomic implications of rising seas will result in difficult choices for coastal development. Federal, state, and local agencies have identified three basic adaptation strategies: allowing the shoreline to retreat and adapting to it; stabilizing the shoreline by erecting walls; and raising the land and nourishing the beach. Public access to the coasts may be reduced by rising seas due to shore loss and shoreline changes. About 50 percent of America's wetlands have al-ready been lost, due mainly to the direct and indirect effects of human activity. In the face of rising seas, current regulatory protection will not achieve the nation's goal of no net loss of coastal wetlands. Existing development of uplands adjacent to wetlands will prevent wetland migration.

The federal government has not taken full advantage of existing programs to develop alternative adaptation strategies for coastal com-munities faced with the effects of sea level rise and global warming. The Council on Environmental Quality, which administers the National Environmental Policy Act, has not required federal agencies to discuss either the effects of global warming and sea level rise or alternative adaptation strategies in their environmental impact assessments.

A number of federal agencies are responsible for coastal policy and management—specifically, the National Oceanic and Atmospheric Ad-ministration (NOAA) in the Department of Commerce, which adminis-ters the Coastal Zone Management Program; the U.S. Army Corps of Engineers; and the Federal Emergency Management Agency (FEMA), which administers the National Flood Insurance Program. There has been little effort, however, to coordinate these agencies' policies and activities or use these programs to reduce the hazards from rising seas. In fact, one agency's programs may conflict with those of other agencies. State coastal zone management programs often discourage the rigid engineering of shoreline stabilization or construction, for example, while the Army Corps of Engineers is both the major proponent and the major builder of such structures. Likewise FEMA—instructed by Congress to discourage development in flood zones—in many cases actually encourages beachfront development by making available inexpensive flood insurance in high-hazard areas.

General Recommendations

Congress should develop effective measures to slow and limit global warming. Congress and the president should adopt a bold program to cut future greenhouse gas emissions drastically in order to avert some of the worst effects of global warming—including the effects of sea level rise and climate change on coastal settlements and ecosystems.

Planners should adopt a basic guideline that takes into account future uncertainties about sea level rise. This guideline should be based on an estimate of sea level rise that has only a 10 to 15 percent probability of being exceeded. Whenever possible, uncertainty about future sea levels should be considered explicitly in the decision-making process. Based on the best available information, the general planning guideline should be a 0.7-meter sea level rise by the year 2050. This number should be revised periodically to reflect new scientific information.

Coastal zoning and management policies should encourage land use adaptation measures such as the siting of new construction—whether public or private—inland of areas likely to be inundated, eroded, or damaged from a sea level rise equal to the general planning assumption (0.7 meter) or from the increasingly frequent and destructive storms that a warming is expected to cause.

Land use adaptation measures are especially critical for new structures that involve substantial public funding or have long use expectancies (such as sewage treatment plants, large industrial facilities, and hazardous waste facilities). These facilities should be located well inland of areas vulnerable to the effects of a 0.7-meter sea level rise—not only to ensure their safety but also to ensure the safety of human settlements and coastal ecosystems that could be polluted by the facilities' disruption during intense storms.

Wherever possible, existing development should be encouraged to relocate landward of areas vulnerable to the effects of a 0.7-meter sea level rise and global warming.

Planners should encourage light recreational uses for areas that could be flooded by a 0.7-meter sea level rise. Public parks and movable structures are examples of appropriate uses.

Current shoreline stabilization projects should avoid nonsustainable rigid solutions, where possible, and should be reviewed in light of the adverse effects from accelerated sea level rise from global warming. In light of the vulnerability of the nation's 95,000 miles of coastline and the potentially massive requests of coastal states for financial assistance, the country should review its financial commitments for coastal stabilization projects.

Recommendations at the Federal Level

- Congress should mandate a nationally coordinated response to sea level rise to ensure that federal investment decisions, public land management, and other federal policies affecting the coasts consider the effects of a 0.7-meter rise in sea level. To this end, Congress should consider the elimination of federal funding of major new facilities and infrastructure located in areas vulnerable to the effects of a 0.7-meter global sea level rise and global warming.
- Congress should mandate that new federal coastal installations, including defense installations, be located inland of areas vulnerable to a 0.7-meter sea level rise unless national security requires otherwise.
- The president should direct the CEQ to require all federal agencies to examine fully the causes and effects of accelerated sea level rise on all major federal actions significantly affecting the quality of the human environment. Specifically, federal agencies should consider the effects of a 0.7-meter sea level rise and increasingly frequent and destructive storms on coastal ecosystems and settlements when determining the feasibility, safety, and environmental effects of federal projects and federally funded, permitted, and licensed activities.
- The U.S. Army Corps of Engineers should take into account the full range of climate change effects when planning its construction and shoreline stabilization projects—including a 0.7-meter sea level rise by 2050, the effects of increasingly frequent and destructive storms on coastal ecosystems, and the impact on the fisheries and marine resources vital to endangered marine species that are expected to be vulnerable to changes in ocean circulation patterns, increased tur-

bidity and water temperature, and the possible loss of spawning and nursery areas. The Corps should also determine how a proposed project will, directly or indirectly, encourage population growth in coastal areas or encourage facilities that might, if flooded, release toxic pollutants into coastal waterways or coastal groundwater. The Corps should review its ongoing and proposed shoreline engineering projects in light of climate change effects on the coasts.

• The National Flood Insurance Program should factor future sea level rise and increasingly frequent and destructive storms from global warming into its risk calculations for purposes of setting insurance rates. Moreover, it should discontinue the practice of issuing flood insurance policies for new development in coastal high-hazard zones. Congress should direct the Federal Emergency Management Agency, which administers the Flood Insurance Program, to determine beach erosion rates using a 0.7-meter rise in global sea level by 2050 and assuming increasingly frequent and intense storms. FEMA also should assess the program's potential financial liability from increased insurance claims related to a 0.7-meter rise in sea level from global warming.

• Congress should use the federally funded Coastal Zone Management Program as an instrument for minimizing the adverse effects of sea level rise and climate change on coastal settlements and ecosystems. Congress should amend the Coastal Zone Management Act (CZMA) to encourage state coastal zone management programs, as part of their federally required improvement tasks, to study, to develop within five years, and to implement management plans addressing the adverse effects on the coastal zone caused by a 0.7-meter sea level rise. These plans should include strategies to: (1) require consideration of global sea level rise in the siting of new infrastructure investments and new large-scale developments that have long life expectancies (sewage treatment plants, industrial plants, hazardous waste facilities); (2) create buffer zones for wetlands that could migrate landward as sea level rises; (3) ensure that structural protection measures are ecologically sound and avoid nonsustainable rigid coastal protection solutions; and (4) require building setbacks and standards that will minimize the adverse effects of sea level rise and storms on human settlements and ecosystems. Congress should increase funding authorization levels by $10 million per year to enable states to carry out these tasks.

- Congress should extend the scope of the permit program under Section 404 of the Clean Water Act. Such an extension would pertain to coastal areas that could become wetlands by the year 2050 from a 0.7-meter sea level rise. Wetlands with migration potential should be identified and buffer zones secured to permit their upland migration.
- Congress should amend the Resource Conservation and Recovery Act (RCRA) to give priority consideration to the immediate cleanup of low-lying coastal hazardous waste facilities and landfills that could be exposed by the increasingly frequent and destructive storms expected from climate change or the effects of a 0.7-meter rise in global sea level by 2050. Congress should prohibit siting of new facilities in such areas.
- Congress should expand the Coastal Barrier Resources System, which prohibits federal development subsidies on undeveloped coastal barriers in the system. As water levels rise, so will the costs of protecting existing structures, the damages from erosion and flooding, the risk to human settlements, and the risk to coastal ecosystems. The federal government should not be subsidizing development that destroys productive coastal ecosystems, endangers the lives and properties of shoreline residents, and costs federal taxpayers millions of dollars each year in flood insurance claims and disaster relief.
- EPA, which initiated sea level rise studies in 1982, should be the lead federal agency for coordinating the efforts of other federal agencies and should continue its policy analysis of coastal effect and alternative adaptation strategies. In addition to coordinating the federal effort, EPA should undertake specific planning tasks. It should reevaluate the strategies in response to accelerated sea level rise, at least every ten years, based on revised global warming trends and increased knowledge about trends in sea level rise. It should conduct beach studies and develop techniques to protect the recreational and conservation values of public lands affected by sea level rise, such as national seashores, beaches, wetlands, and barrier islands. It should calculate the funds needed to purchase easements on lands adjacent to public beaches capable of inland migration. It should assess the feasibility of a federal system of sand rights ensuring that stabilization measures undertaken at up-current locations will not rob down-current shorelines of sand or otherwise endanger them. Finally, EPA should develop a program to help coastal communities assess the effect of sea level rise and climate change on

their drinking water supplies, including aquifers. Congress should provide funding to help these communities redesign their water systems where necessary.

- Congress should direct the National Ocean Service (NOS) to collect data from global positioning satellites and analyze tide gauge data worldwide to predict future trends. The NOS should also conduct an analysis of the effects of a 0.7-meter global sea level rise by 2050 and the impact of global warming.
- Congress should expand research on projections of future sea level rise, based on general circulation models, and the critical processes that influence coastal response to sea level rise. Research on the responses of coastal wetlands to sea level rise is particularly critical.
- Congress should direct the U.S. Geodetic Survey (USGS) to map 2-foot contours for delineation of the present and future 100-year floodplain to assist in the development of a coastal hazards data base. Some areas on present USGS topographic maps have only 20-foot contours, some barrier islands on the USGS quadrangle map show no elevation at all, and some topographic maps were last updated in the 1940s.
- Congress should direct the National Science Foundation (NSF) to support basic research in coastal erosional processes and the connection between sea level positions and past climates. Development of a Global Coastal Hazards Data Base should be encouraged. Such a data base would enable, in turn, the development of a Coastal Vulnerability Index that could predict the coastal segments at greatest risk to a rise in sea level caused by future climate warming. To date, NSF has provided insufficient funding for coastal studies.
- Congress should revise federal tax policies to ensure that they do not encourage unwise investment in areas vulnerable to the effects of sea level rise and global warming or investment in projects that might prove destructive to coastal ecosystems. In particular, casualty loss deductions and preferential treatment of industrial development bonds to finance infrastructure should be phased out in such areas. (Industrial development bonds are tax-exempt state and local bonds used to finance development like roads, bridges, and other infrastructure.) Congress should eliminate the tax exemption on income from such industrial development bonds.
- Consistent with the Climate Institute's recent recommendations, Congress should appropriate at least $25 million (in 1989 dollars) to conduct necessary research and policy preparation for adaptation to sea level rise from global warming.

Recommendations at the State Level

- All coastal states should initiate studies of the anticipated effects of a 0.7-meter rise in sea level by the year 2050. These studies should estimate the effects of worst-case scenarios—especially the effects of storm surge in combination with a 0.7-meter sea level rise—and identify the low-lying coastal ecosystems and human settlements most vulnerable to rising global sea levels. The list of threatened areas and resources should provide the basis for state land use adaptation and coastal protection measures that are ecologically sound.
- All coastal states should establish setback requirements to ensure the safety of structures from a 0.7-meter sea level rise by 2050 and from increasingly frequent and intense storms.
- Structures that are vulnerable to erosion, flooding, and storm damage from a 0.7-meter sea level rise should be relocated landward whenever possible. Rebuilding of structures significantly damaged (50 percent or more) or destroyed by erosion or storm flooding should be prohibited in such areas.
- Coastal permitting and procedures should take into account the effects of a future global sea level rise of 0.7 meter and climate change. New building permits should not be issued unless the applicant can present convincing evidence that the building's design and location are appropriate in light of the effects of a 0.7-meter sea level rise by 2050 and climate change, including the potential effect of the building on coastal ecosystems. Generally, new building permits should be available only for movable structures.
- As a general rule, states should prefer coastal protection methods such as beach nourishment over nonsustainable rigid control methods such as seawalls. Priority should be given to methods that minimize effects on adjacent properties and ecosystems. In general, hard structural methods are appropriate only for highly developed, urban areas. Groins, jetties, and similar structures should be used only at the terminal end of the sand supply.
- State environmental quality review statutes should require consideration of the effects of a 0.7-meter sea level rise and global warming in environmental impact statements.
- Wetland protection statutes and regulations should be modified to provide for a sufficient buffer to allow wetland migration landward as sea levels rise.

- States should analyze potential salinization and contamination of aquifers and groundwater from the effects of a 0.7-meter sea level rise and climate change. They should plan for adequate drinking water supplies and sufficient sewage and waste treatment facilities. Efficient water supply management, water conservation, and reservation of lands for future reservoir needs should be a priority.
- State statutes and regulations governing hazardous waste facility siting should be modified to prohibit location of these facilities in the coastal zone and to require proof (engineering certainty) that new facilities will not be affected by a 0.7-meter sea level rise or by increasingly frequent and intense coastal storms. Plans should also be made for closing, relocating, and securing existing facilities that may be threatened by the effects of a 0.7-meter sea level rise and climate change.
- States should initiate public education regarding the effects of global warming and sea level rise on coastal communities and ecosystems.
- States with federally approved coastal zone management programs should ensure that all federal activities in their coastal zone are consistent with state policies relating to sea level rise and global warming.

Appendix A:

The 1989 Cairo Compact

Statement and Resolutions of Working Group 3 Relating to Coastal Areas, Fisheries, and Marine Resources

There is now scientific consensus that increased atmospheric concentrations of carbon dioxide and other greenhouse gases are likely to lead to changes in climate and that this could occur in the next few decades.

A warming of the atmosphere by several degrees is anticipated with a slower warming of the oceans. This will bring with it changes to the general circulation of the global atmosphere and oceans and changes to the distribution of the climatic patterns of the world.

Current sea-level rise is of the order of 0.1 to 0.25 meter per century. An accelerated rise in sea levels is anticipated. Current model calculations suggest that sea level will rise 0.3 m ± 0.2 m by 2030 if the atmosphere warms by 3°C. Sea-level rise will continue and possibly accelerate beyond 2050. However, such global averages mask the substantial variability that will occur in different coastal regions. The frequency and effect of extreme effects such as storm surges and other intensified perturbations will also create a greater uncertainty.

The political and economic infrastructure of most societies has developed with an underlying assumption of a variable climatic environment which is stable about a long-term mean. The greenhouse warming of the atmosphere undermines this assumption of long-term stability and thus, as well as environmental changes occurring, substantial social and economic change will also ensue.

Our current state of knowledge of those coastal areas most at risk is inadequate, and this will become a priority in the development of response options to sea-level rise and other impacts. Nations with low-lying deltaic areas and island states will be especially vulnerable. Severe economic and social dislocation can be expected for present and future generations as a large and growing proportion of the world's population, development, and facilities are based on the coasts.

Whilst it is anticipated that most major cities in both industrialized and developing countries will develop coastal defense structures for protection against sea-level rise, this will be at great expense. As a result, serious adverse impacts can be expected on economic development in many areas. A range of new opportunities will also arise, but the scope and nature of these are difficult to predict.

In developing countries, the impact of sea-level rise will be most severe on exposed coastal populations and on agricultural developments in low deltaic areas. Countries with heavily populated coastal areas would be particularly susceptible to rising sea level because of the delicate physical and ecological balance achieved by deltas at the river/ocean interface, local subsidence, the erosional forces of oceanic energy, waves and storms generated by extreme events, and changes to river systems. Salination of many freshwater systems can be expected. Millions of people would be displaced and a significant portion of arable land lost if current predictions of climate change are realized. For example, a one-meter sea-level rise might involve the displacement of 300 million people.

It is considered that the most extreme situation will be faced by island states, which will be particularly vulnerable to rising sea levels and storm-generated surges. More frequent and intense storms would cause extensive damage to the land, infrastructure, and freshwater supplies and even threaten the survival of some island states. Their physical location provides no "retreat" option; they have limited resources and infrastructure for the "defend" option, especially with small populations spread over many tiny islands. Their economies are already largely dependent on aid, migration, and income from nationals working overseas.

Rising sea levels will lead to the loss of coastal land uses, including agriculture, ports, industrial zones, aquaculture, and coastal protection systems such as mangroves and coral reefs.

For some nations the only solution available in responding to rising sea levels and increased storm surges and climate extremes may be abandonment. The social upheaval and economic impact of such actions may have serious repercussions for international security and the world economy. Such changes would have socially and economically devastating impacts on the nation itself.

Fisheries and marine resource production are expected to be vulnerable to global change, including increased exposure to solar ultraviolet radiation and sea-level rise plus other greenhouse-induced effects such as changes to ocean circulation patterns, increased turbidity, water temperature, and storminess, and the possible loss of spawning and nursery areas, especially mangroves and seagrass flats.

Today, the living resources in many areas of the sea are under threat from overexploitation, pollution, and land-based development. Most major familiar fish stocks throughout the waters covering the continental shelves, which provide 95 percent of the world's fish catch, are now threatened by overexploitation.

For many fisheries there is insufficient knowledge of the dynamics of fish populations to manage them as a resource stock. However, recent studies have suggested a strong influence of climate factors on recruitment. Effects on mangroves, coral reefs, coastal wetlands, other habitats, and nursery grounds have been shown to arise from climate-related factors. There is little doubt that a rapidly changing marine environment due to global warming and stratospheric ozone depletion will have an influence on the distribution, availability, and types of living marine resources and marine ecosystems.

Pressures will be for the development of measures to mitigate overexploitation of marine food resources. The instigation of international agreements for the management of most major fisheries, intensified research, and the development of inventories and monitoring programs for enhanced stock management will be required. The development of alternative sources of food, employment, and income will be necessary for those people in locations susceptible to adverse impacts of climate change.

There could be both positive and negative effects on mariculture. Many mariculture systems benefit from warmer water temperatures. However, the fixed nature of mariculture could also lead to production losses.

The conference recognizes the work of the Intergovernmental Panel on Climate Change (IPCC), which was established by UNEP and WMO and recognized by the UN General Assembly Resolution 43/53 on Protection of Global Climate for Present and Future Generations of Mankind.

The conference recognizes climate change as a global concern that threatens the well-being of present and future generations. Furthermore, the conference recognizes the need to stabilize CO_2 emissions and emissions of other greenhouse gases not controlled by the Montreal

Protocol. The conference considers that any delay in taking action to halt the progressive deterioration in environmental stability will contribute further to the social and economic dislocation already expected.

A Call for Action

In the light of the scientific consensus regarding the likelihood of climate change and global warming, and deeply concerned over the changing global environment and its possible adverse effects, particularly the threat of sea-level rise and increased frequency and intensity of many natural disasters, this conference resolved to act collectively and individually to the following program of action, which is designed to address the impacts of climate change and global warming on coastal areas, fisheries, and marine resources.

1. Urges immediate action by governments, the United Nations and its specialized agencies, other international bodies, nongovernment organizations, industry, educational institutions, and individuals to establish a global plan to achieve stabilization of atmospheric concentrations of greenhouse gases.

Coastal Zone Management

2. Recommends that all nations establish, adapt, and strengthen institutions to protect and manage their coastal zones, so that they may anticipate and plan effectively for sea-level rise and increased intensity and frequency of extreme meteorological events.
3. Recommends that all nations give priority to land-use adaptation and ecologically sound protection measures avoiding where possible nonsustainable rigid solutions.
4. Recommends that all nations identify and protect (to the limit of the means at their disposal) ecosystems that are essential to the sustainable use of living resources on which communities depend for food and industry, to preserve genetic resources required for the development and improvement of domestic plants and animals

and for the advance of science and technology, and to maintain essential ecological processes and life-support systems on which human survival and development depend.

5. Recommends that all nations adopt a proactive approach, including environmental impact assessment, to protect coral reefs, salt marshes, mangroves, and coastal wetlands according to their individual requirements and capabilities in recognition of the important role they play in forming barriers against sudden sea surges, trapping sediments, and protecting the coast behind them.

6. Calls upon all nations to undertake environmental assessment studies for all development projects, to review existing development programs to ensure that likely greenhouse impacts are considered, and to aid the development of indigenous capabilities for environmental assessments where this does not already exist.

7. Recommends that all nations develop and contribute to national and regional planning strategies to prevent development that would be seriously at risk under anticipated changes to sea level and climate. In particular, calls upon national planning agencies to ensure that sea-level rise and related climatic impacts are taken into account when new developments in coastal areas are being considered.

8. Recommends that all nations adopt measures to discourage migration to, and limit population growth in, vulnerable coastal areas.

Fisheries Management

9. Recommends that strategies and data bases be developed on national and international levels to facilitate improved fisheries management and protection of marine living resources, including mariculture, in light of the additional stresses due to climate change and stratospheric ozone depletion. This should be done using established scientific approaches and taking into account existing national and international legal frameworks.

10. Recommends that successful demonstration programs in aquaculture and mariculture be identified and every effort be made to provide practical examples of the economic benefits that can be derived from such ventures.

Appendix B:

A Draft Protocol for International Cooperation

International agreement on a treaty to control greenhouse gas emissions may depend on the industrialized nations' willingness to cooperate and assist less developed nations in responding to the adverse effects of sea level rise and global climate change. The following draft protocol, which could be adopted at the same time as an international treaty to control greenhouse gases, provides a mechanism for such cooperation and assistance among nations.

The contracting parties to this protocol,

Being contracting parties to the Framework Convention for the Protection of the Climate done _____

on _____,

Conscious that global climate change will cause significant global sea level rise, as presented by Working Group 1 of the Intergovernmental Panel on Climate Change, will result in increased intensity and frequency of extreme meteorological events and storms, and will aggravate existing relative sea level rise from subsidence and other localized conditions,

Recognizing that global cooperative research and monitoring programs are necessary for the effective prediction and response to the effects of sea level rise and climate change,

Aware that nations with low-lying deltaic areas and island states will be especially vulnerable and that severe economic and social dislocation can be expected for present and future generations as a large and growing proportion of the world's population, development, and facilities are based on the coasts,

Recognizing that rising sea levels will lead to the loss of coastal land uses and ecosystems, including estuaries and coastal protection systems such as mangroves, coral reefs, and coastal barriers, ports, coastal agriculture, and other critical habitats,

Also recognizing that fisheries and marine resources vital to endangered marine species are expected to be vulnerable to global change-induced effects, such as changes in ocean circulation patterns, as well as increased turbidity, water temperature, and storm activity, and to the possible loss of spawning and nursery areas,

Aware that rising sea levels will cause increased salinization of coastal rivers and groundwater sources and increase the opportunity for release of hazardous substance due to potential flooding of low-lying hazardous waste sites or solid waste disposal facilities, especially from storm surge during major coastal storms,

Recognizing the importance of sound preparation, cooperation, and financial and technical assistance in responding to these threats in order to minimize damage to coastal resources and human settlements,

Acknowledging that near-term measures, such as directing new development and population away from areas vulnerable to flooding and erosion and taking steps to preserve upland buffers to wetlands, can substantially reduce many adverse effects of sea level rise,

Noting that the United Nations Convention on the Law of the Sea acknowledges that landlocked states have rights and interests in the exploitation and conservation of living marine resources,

Have agreed as follows:

Article 1

DEFINITIONS

Article 2

APPLICATIONS

This protocol applies to all parties to the Convention for the Protection of the Climate, including both landlocked states and those whose coastal zones are subject to the adverse effects of sea level rise and global climate change.

Article 3

GENERAL PROVISIONS

1. The contracting parties shall, within their capabilities, cooperate in taking all necessary planning measures to protect coastal resources

and human settlements from the adverse effects of sea level rise and global climate change.

2. The contracting parties shall, within their capabilities, establish, maintain, and provide financial and technical assistance for anticipating and responding to the threat of sea level rise and global climate change to coastal communities and ecosystems. Such means shall include:

   a. Establishing, adapting, and strengthening institutions to protect and manage their coastal zones, with particular emphasis on sea level rise and climate change

   b. Giving priority to land use adaptation and ecologically sound protection measures, avoiding where possible nonsustainable rigid solutions

   c. Identifying and protecting ecosystems that are essential to the sustainable use of living resources on which communities depend for food and industry

   d. Adopting a proactive approach, including the preparation of environmental impact assessments, for protection of coral reefs, salt marshes, mangroves, coastal marshes, and coastal barriers

3. Each contracting party shall periodically provide to the secretariat systematic information relating to the implementation of this protocol including information on rates of sea level rise and appropriate adaptive strategies.

Article 4

MUTUAL ASSISTANCE

1. The contracting parties will cooperate to strengthen the ability of international institutions to anticipate the potential effects of sea level rise, to promote information exchange, and to assist in planning for sea level rise. Information dissemination tasks will be assigned to the secretariat established by Article 9 of this protocol.

2. Each contracting party shall render assistance, within its capabilities, to other contracting parties that request assistance in responding to sea level rise and global climate change.

3. Each contracting party will seek, locate, identify, or establish means for providing financial and technical assistance to other contracting parties to assist in developing indigenous strategies for combating the effects of sea level rise and global climate change. This effort could include working with existing multilateral devel-

opment banks and international aid agencies, as well as establishing new funding entities for the purposes of this protocol.

Article 5

RESEARCH AND MONITORING

1. The contracting parties will cooperate on international research into the amount and timing of global and regional sea level rise, as well as the frequency and intensity of storms and other atmospheric/oceanic phenomena.
2. The contracting parties will cooperate on an international and regional basis on research programs to understand the economic, social, scientific, and environmental effects of sea level rise and global climate change.
3. The contracting parties will cooperate on the development of strategies and data bases on national and international levels to facilitate improved fisheries management and protection of marine living resources, in light of the additional stresses due to sea level rise and global climate change.

Article 6

PUBLIC EDUCATION AND PARTICIPATION

1. Each contracting party shall facilitate education at the local and national level to develop awareness of the risks to coastal areas, fisheries, and marine resources from existing activities and future sea level rise.
2. Each contracting party, to the extent consistent with its own legal institutions, will ensure adequate public participation in decisions affecting the development and protection of coastal areas, fisheries, and marine resources.

Article 7

SHARED COASTAL RESOURCES
AND SETTLEMENTS OF DISPUTE

Each contracting party agrees to submit conflicts over shared coastal resources, such as shared coastlines, estuaries, and fisheries, in accordance with procedures in Article _____ of the Convention.

Article 8

ADOPTION AND AMENDMENT OF THE PROTOCOL

[To be consistent with provisions developed within the body of the Framework Convention.]

Article 9

INSTITUTIONAL ARRANGEMENTS

The contracting parties designate the United Nations Environment Program to carry out the following secretariat functions:

1. Collect and disseminate information on sea level rise and adaptive strategies.
2. Prepare and convene meetings of the contracting parties.
3. Coordinate the implementation of cooperative activities agreed upon by the meetings of contracting parties and conferences.

Article 10

EMERGENCY FACILITIES

The contracting parties shall establish an emergency facility to mobilize timely and effective support for areas stricken by sea level rise and climate change in order to alleviate human suffering and assist communities in coping with adverse effects, including flooding. [This provision on emergency facilities may also be included in the Convention.]

Article 11

REGIONAL SEA PROGRAMS

The contracting parties are encouraged to develop shared coastal management plans for shared regional seas and water bodies. Such coastal management plans shall be consistent with this protocol.

Appendix C:

CZMP Responses to Sea Level Rise

State	Recognition of Problems and Issues by CZMP	New Public and Inter-governmental Processes	Existing Adaptable Regulation	New Policies Responding to Sea Level Rise
Alabama	no	no	partial	no
Alaska	no	no	no	no
California	yes	no	no	no
SFBCDC[a]	yes	yes	yes	yes
Connecticut	no	no	no	no
Delaware	yes	yes	partial	no
Florida	yes	no	partial	no
Georgia	no	no	no	no
Hawaii	yes	no	no	no
Louisiana	yes	yes	partial	no
Maine	yes	yes	yes	yes
Maryland	yes	yes	partial	no
Massachusetts	yes	no	no	no
Mississippi	no	no	no	no
New Hampshire	yes	no	no	no
New Jersey	yes	yes	partial	no
New York	yes	yes	partial	no
North Carolina	yes	yes	yes	no
Oregon	yes	yes	no	no
Pennsylvania	no	no	no	no
Rhode Island	yes	no	partial	no
South Carolina	yes	yes	yes	yes
Texas	no	no	partial	no
Virginia	no	no	no	no
Washington	yes	yes	no	no

NOTE: The Great Lakes states, Puerto Rico, Virgin Islands, Northern Marianas, American Samoa, and Guam are not included in this table. "Partial" denotes that these policies impose partial restrictions on coastal development.

[a] Regional CZMP authority having limited jurisdiction in California.

SOURCE: *Klarin and Hershman (1990). Reprinted with the authors' permission.*

Appendix D:

Alternative Policy Responses to Sea Level Rise

Range of Policy Responses	Examples of Federal or State Policies
Erosion-based setbacks	South Carolina Beach Management Act: establishes setbacks 40 times annual erosion rate. The baseline for the setback is reset every 5 to 10 years.
	North Carolina Coastal Areas Management Act: establishes annual erosion rate setbacks of 30 times and 60 times for single and multiple residences.
Building codes and size restrictions	Maine Sand Dune Law: new development restricted to 2,500 sq ft and 35 ft height. Single and multiple residence buildings must be 1 ft and 4 ft above base flood elevation in low hazard zones.
	North Carolina Coastal Areas Management Act: restricts structure size to 5,000 sq ft in shoreline setback areas.
Development restrictions in flood hazard areas	Maine Sand Dune Law: new development restricted to low hazard areas not to exceed 40% of undeveloped dune areas, with 20% being buildings.
	Florida Construction Control Lines: establish area within which new development must be permitted. No construction within 30-year erosion zone.
Economic incentives/disincentives	
Restrict new infrastructure and flood insurance availability	Coastal Barrier Resource Act: no federal subsidies for infrastructure or flood insurance within Coastal Barrier Resource System.
Incentives to remove or relocate structures upland	Upton-Jones Act: Federal Flood Insurance Program pays owners to remove or relocate damaged structures in hazardous flood zones.
Proposed tax incentives to control development	Delaware Beaches 2000 Plan: proposes favorable tax assessments to property owners who develop property for uses compatible with preservation of beaches.

Range of Policy Responses	Examples of Federal or State Policies
	Project planning
Engineering standards	San Francisco Bay Conservation and Development Commission: Bay Plan requires proposed developments to consider sea level rise in project engineering plans under the review process.
Remodel or redesign infrastructure	Charleston, South Carolina: designed new flood control and drainage system to account for sea level rise and subsidence over next 50 years.
	Prohibit or restrict development
Post-storm reconstruction restrictions	South Carolina Beach Management Act: restrictions on reconstruction of structures destroyed in excess of 66% by storms within setback zones. Replace all vertical erosion and protection structures over 30-year period.
	Texas: Open Beaches Act prohibits reconstruction of damaged buildings and protective devices on property seaward of the vegetation that is open to public access.
Land acquisition and conservatory programs	California Coastal Conservatory uses state bond monies to acquire undeveloped coastal property.
	Florida buys property for preserving public beaches, public access, and recreation areas.
Preserve critical habitats and wetlands	Maryland Chesapeake Bay Critical Areas Act: establishes buffer around wetlands and reduces density of adjacent development.
Proposed abandonment policy for coastal areas	Long Island Regional Planning Board Proposal to end long-term leases of state coastal property and buyback of barrier island properties severly damaged by storm floods.
	Non-structural engineering
Resedimentation of river deltas	Louisiana Coastal Environmental Protection Trust Fund and State/Federal Joint Task Force: local resedimentation projects.
Beach renourishment, dune and wetlands revegetation, and stabilization programs	Florida: Beach Management Fund authorizes up to $35 million annually toward beach erosion, preservation, and restoration projects.
	Maryland: $60 million multiyear federal, state, and local plan to renourish ocean beach shoreline.
	South Carolina: BMA requires property owners to replenish sand at 150% of annual volume to replace destroyed structural erosion devices.
	Groundwater protection
Preserve coastal aquifers and groundwater resources	Maine: requires permit applications be reviewed by District Water Company to determine impact on groundwater recharge.

SOURCE: *Klarin and Hershman (1990). Reprinted with the authors' permission.*

Excerpts from Maine's Sand Dune Rules

Summary

These rules clarify the criteria for obtaining a permit under Maine's Sand Dune Law (38 M.R.S.A., Sections 471–478). The rules outline classes of projects that are exempted from the requirement of obtaining a permit. For all other projects, the rules outline standards that projects must meet in order to satisfy the statutory criteria. The rules also contain a definitions section, and a section which provides general information on the processing of sand dune applications.

Section 3: Standards

PREAMBLE

The Board recognizes that coastal sand dunes change over time due to the forces of wind and waves on the sand. Evidence exists that sea level is currently rising. In addition, theories have been developed which predict this rise to accelerate in the future. This rise will increase the rate of shoreline erosion and flooding, and the risk of damage to coastal property. Historical evidence has shown that attempts to prevent erosion and flooding through the construction or enlargement of seawalls result in harm to the beach for primarily two reasons:

1. Seawalls reflect waves onto the beach causing sand to be scoured away.
2. Seawalls cut off the natural supply of sand to the beach from the sand dune area behind the wall.

Usually, under these circumstances, a beach can only be maintained by continually adding sand to the beach. Such beach nourishment programs result in continually increasing costs to both the public and private sectors for maintenance of the beaches and adjacent development.

The extent to which sea level will change in the future is uncertain. However, under any scenario of increasing sea level, the extensive development of sand dune areas and the construction of structures which are not practical to move increase the risk of harm, both to the sand dune system and to the structures themselves.

Therefore, in order to protect the natural supply and movement of sand, and to prevent creation of flood hazards, the Board will evaluate proposed developments with consideration given to future sea level rise and will impose restrictions on the density and location of development and on the size of structures.

STANDARDS

Activities in a sand dune system must meet the minimum standards set out in the following paragraphs. . . .

A. All Projects

1. Projects shall have a minimal impact on the immediate site and on the sand dune system. Impacts which may reasonably be expected to occur during the following 100 years will be considered.
2. Projects shall not be permitted if, within 100 years, the project may reasonably be expected to be damaged as a result of changes in the shoreline. Beach nourishment and dune construction projects are excluded from this requirement.
3. Projects shall not cause a flood hazard to any structure during a 100-year flood or storm.
4. Shorebird nesting or breeding areas or activities shall not be unreasonably disturbed by any project activities. Shorebird nesting or

breeding areas shall be adequately buffered from subsequent human activities associated with the use of any project. Buffer requirements will be based upon the best available data.

5. Projects shall not unreasonably interfere with legal access to or use of the public resources.
6. Disturbed areas of natural beach vegetation shall be restored as quickly as possible. Natural beach vegetation includes American beach grass, rugosa rose, bayberry, beach pea, beach heather, and pitch pine.

B. Structures
1. All Structures
   The approval of all new, reconstructed, and replaced structures, except for piers, and additions, a combined total of which cover less than 250 square feet of ground surface since the effective date of these rules, shall be subject to the following conditions:
   a. No seawall shall be constructed or expanded on the property.
   b. If the shoreline recedes such that the coastal wetland as defined under 38 M.R.S.A. Section 472 extends to any part of the structure, including support posts, for a period of six months or more, then the approved structure, along with appurtenant facilities, shall be removed and the site shall be restored to natural conditions within one year.
   c. Any debris or other remains from damaged structures on the property shall be removed from the sand dune system.
   d. No structure shall be relocated within the sand dune system without approval of the Maine Department of Environmental Protection.

Note: The conditions in subparagraphs (a) through (d) are based on the conclusions of the Department that:

- Sea level is rising, and the amount of shoreline erosion and frequency of flooding will increase.
- Seawalls interfere with the supply and movement of sand and accelerate beach erosion.
- Structures which are located in a coastal wetland interfere with the natural supply and movement of sand within the sand dune system and create an unreasonable flood hazard.

NOTES

CHAPTER 1

1. T.M.L. Wigley, P. D. Jones, and P. M. Kelly, "Empirical Climate Studies: Warm World Scenarios and the Detection of Climatic Change Induced by Radioactively Active Gases," in *The Greenhouse Effect, Climatic Change, and Ecosystems*, ed. B. Bolin et al. (New York: Wiley, 1986), pp. 271–311.
2. World Meteorological Organization/United Nations Environment Programme: Intergovernmental Panel on Climate Change, *Climate Change: The IPCC Scientific Assessment*, ed. J. T. Houghton, G. J. Jenkins, and J. J. Ephraums (Cambridge, England: Cambridge University Press, 1990) p. 1 of Executive Summary.
3. W. S. Broecker, "Unpleasant Surprises in the Greenhouse?" *Nature* 328 (1987):123–126.
4. D. Lashof, "The Dynamic Greenhouse: Feedback Processes That May Influence Future Concentrations of Atmospheric Trace Gases and Climatic Change," *Climatic Change* 14 (1990):213–242.
5. R. A. Kerr, "New Greenhouse Report Puts down Dissenters," *Science* (1990):481–482.
6. Wigley et al., "Empirical Climate Studies," p. 277.
7. R. E. Dickinson and R. J. Cicerone, "Future Global Warming from Atmospheric Trace Gases," *Nature* 319 (1986):109–115.
8. J. E. Hansen et al., "Climate Sensitivity to Increasing Greenhouse Gases," in *Greenhouse Effect and Sea Level Rise: A Challenge for This Generation*, ed. M. C. Barth and J. G. Titus (New York: Van Nostrand Reinhold, 1986), pp. 57–77.
9. J. Hansen, A. Lacis, and M. Prather, "Greenhouse Effect of Chlorofluorocarbons and Other Trace Gases," *Journal of Geophysical Research* 94 (November 1989):16,417–16,421.
10. R. H. Gammon et al., "History of Carbon Dioxide in the Atmosphere," in *Atmospheric Carbon Dioxide and the Global Carbon Cycle* (Washington, D.C.: Department of Energy, 1985).
11. See G. Woodwell, "Biotic Effects on the Concentration of Atmospheric Carbon Dioxide: A Review and a Projection," in *Changing Climate* (Washington, D.C.: National Academy Press, 1983). See also G. Woodwell et al., "Global Deforestation: Contribution to Atmospheric Carbon Dioxide," *Science* 222 (1983):1081–1086.

12. Hansen et al., "Climate Sensitivity," pp. 55–77.

13. M.K.W. Ko and N. D. Sze, "A 2-D Model Calculation of Atmospheric Lifetimes for N_2O, CFC-11, CFC-12," *Nature* 287 (1982):317–319; see also D. M. Cunnold et al., "The Atmospheric Lifetime Experiments 3 and 4: Lifetime Methodology and Application to Three Years of CFC Data," *Journal of Geophysical Research* 88 (1983):8379–8414.

14. National Academy of Sciences, *Changing Climate* (Washington, D.C.: National Academy Press, 1983), p. 24.

15. D. Doniger and D. Wirth, "Cooling the Chemical Summer," in *Effects of Changes in Stratospheric Ozone and Global Climate*, vol. 1 (New York: U.N. Environmental Program/U.S. Environmental Protection Agency, 1986), p. 345.

16. P. J. Crutzen and L. T. Gidel, "A Two-Dimensional Photochemical Model of the Atmosphere 2: The Tropospheric Budget of the Anthropogenic Chlorocarbons, CO, CH_4, CHCl and the Effect of Various NO_x Sources on Tropospheric Ozone," *Journal of Geophysical Research* 88 (1983):6641–6661; see also testimony of V. Ramanathan before the Senate Subcommittee on Environmental Protection, on "The Greenhouse Effect and Climate Change," January 23, 1987.

17. R. Kerr, "Is the Greenhouse Here?" *Science* 239 (1988):559.

18. R. Kerr, "Global Warming Continues in 1989," *Science* 343 (1990):521.

19. A. H. Lachenbruch and B. V. Marshall, "Changing Climate: Geothermal Evidence from Permafrost in the Alaskan Arctic," *Science* 234 (1986):689–696.

20. "Warming of Alaskan Arctic Seen as Evidence of Greenhouse Effect," *New York Times*, October 31, 1986.

21. V. Ramanathan, "Trace Gas Trends and Change," testimony before the Senate Subcommittee on Environmental Protection, January 23, 1987.

22. J. Hansen et al., "The Greenhouse Effect: Projections of Global Climate Change," in *Effects of Changes in Stratospheric Ozone and Global Climate*, vol. 1 (New York: UNEP/EPA, 1986), p. 199.

23. J. Hansen et al., "Global Climate Changes as Forecast by the GISS 3-D Model," *Journal of Geophysical Research* 93 (1988):9341–9364.

24. I. M. Mintzer, *A Matter of Degrees: The Potential for Controlling the Greenhouse Effect* (Washington, D.C.: World Resources Institute, 1987), p. 39.

25. The Advisory Group for Greenhouse Gases, *Responding to Climate Change: Tools for Policy Development (Summary Report)*, ed. J. Jäger (Stockholm, Sweden: The Stockholm Environment Institute, 1990), p. 4.

26. Ibid., p. 5.

CHAPTER 2

1. P. W. Glynn, "Widespread Coral Mortality and the 1982–83 El Niño Warming Event," *Environmental Conservation* 11 (2) (1988):133–146.

2. V. Gornitz, S. Lebedeff, and J. Hansen, "Global Sea Level Trend in the Past Century," *Science* 215 (1982):1611–1614.

3. P. Vellinga and S. Leatherman, "Sea Level Rise, Consequences and Policies," *Climate Change* 15 (1989):175–189.

4. See J. Hoffman, J. Wells, and J. Titus, *Projecting Future Sea Level Rise* (Washington, D.C.: EPA, 1983); see also R. Revelle, "Probable Changes in Sea Level Resulting from Increased Atmospheric Carbon Dioxide," in National Academy of Sciences, *Changing Climate* (Washington, D.C.: National Academy Press, 1983), pp. 433–444. (Revelle excluded the ice cap contribution from his sea level rise prediction of 70 cm ± 18 cm over the next century. He noted, however, that sea level could rise 1 to 2 meters (39–79 inches) per 100 years in the following centuries if portions of the Antarctic ice cap were to melt completely.)

5. J. Smith and D. Tirpak (eds.), *The Potential Effects of Global Climate Change on the United States*, report to Congress (Washington, D.C.: EPA, 1989), p. 123.

6. National Research Council, *Responding to Changes in Sea Level: Engineering Implications* (Washington, D.C.: National Academy Press, 1987).

7. R. B. Alley and I. M. Williams, "The Unsteady State of the West Antarctic Ice Sheet," paper presented at the NSF Sea Rise Workshop, January 23–25, 1990.

8. M. F. Meier, "News and Views: Reduced Rise in Sea Level," *Nature* 343 (1990):115–116.

9. Intergovernmental Panel on Climatic Change, "Adaptive Options and Policy Implications of Sea Level Rise and Other Coastal Impacts of Global Climatic Change," workshop report to the Coastal Zone Management Subgroup, November 27–December 1, 1989, Miami, Florida.

10. H. Zwally, A. Brenner, J. Major, R. Bindschadler, and J. Marsh, "Growth of Greenland Ice Sheet: Measurement," *Science* 246 (1990):1587–1591.

11. M. Kuhn, "Small Glacier Wastage," paper presented to the American Geophysical Union, October 24, 1989.

12. Conversation between Eric Washburn and Dr. David Bromwich, Ohio State University Polar Studies Institute, at the NSF Sea Rise Workshop, January 24, 1990.

13. Telephone conversation with Mark Meier, January 8, 1990.

14. Conversation between Eric Washburn and Dr. David Bromwich, January 23, 1990; conversation between Eric Washburn and Dr. Doug MacAyeal, January 22, 1990.

15. P. Vellinga and S. Leatherman, "Sea Level Rise, Consequences and Policies," *Climatic Change* 15 (1989):175–189.

CHAPTER 3

1. P. Bruun, "Worldwide Impacts of Sea Level Rise on Shorelines," in *Effects of Changes in Stratospheric Ozone and Global Climate*, vol. 4 (New York: UNEP/EPA, 1986), pp. 99–128; P. Bruun, "Sea Level Rise as a Cause of Shore Erosion," *Proceedings of the American Society of Engineers and Journal Waterways Harbors Division* 88 (1962):117–130.

2. R. Sorensen et al., "Control of Erosion, Inundation, and Salinity Intrusion Caused by Sea Level Rise: A Challenge for This Generation," in *Greenhouse Effect and Sea Level Rise*, ed. M. C. Barth and J. G. Titus (New York: Van Nostrand Reinhold, 1984), p. 179.

3. S. Leatherman, "Barrier Island Evolution in Response to Sea Rise: A Discussion," *Journal of Sedimentary Petrology* 53 (1983):1026–1031.

4. G. Griggs and L. Savoy (eds.), *Living with the California Coast* (Durham: North Carolina University Press, 1985).

5. California Coastal Commission, *Draft Report on the Effects of Sea Level Rise on California's Coasts*, 1989.

6. D. deSylva, "Increased Storms and Estuarine Salinity and Other Ecological Impacts of the Greenhouse Effect," in *Effects of Changes in Stratospheric Ozone and Global Climate*, vol. 4 (New York: UNEP/EPA, 1986), pp. 153–164.

7. P. A. Douglas and R. H. Stroud, *A Symposium on the Biological Significance of Estuaries* (Washington, D.C.: Sport Fishing Institute, 1971), p. 7.

8. R. Park, T. Armentano, and C. Cloonan, "Predicting the Effects of Sea Level Rise on Coastal Wetlands," in *Effects of Changes in Stratospheric Ozone and Global Climate*, vol. 4 (New York: UNEP/EPA, 1986), pp. 129–152.

9. T. Kana et al., *Potential Impacts of Sea Level Rise on Wetlands Around Charleston, South Carolina* (Washington, D.C.: EPA, 1984).

10. Kana et al., *Potential Impacts*.

11. EPA and Louisiana Geological Survey, "Saving Louisiana's Coastal Wetlands: The Need for a Long-Term Plan of Action," EPA-230-02-87-026 (Washington, D.C.: EPA, 1987).

12. F. Gibney, "Louisiana's Bayou Blues," *Newsweek*, June 22, 1987.

13. P. W. Glynn, "Widespread Coral Mortality and the 1982–83 El Niño Warming Event," *Environmental Conservation* 11 (2) (1988):133–146.

14. National Council on Public Works Improvement, *Fragile Foundations: A Report on America's Public Works*, final report to the president and Congress (Washington, D.C.: 1988).

15. W. Hyman, *Potential Impacts of the Greenhouse Effect on the Infrastructure of Greater Miami, Florida* (Washington, D.C.: Urban Institute, 1988).

16. Conversation with Christopher Cole, August, 1988.

17. G. Shih, "Everglades National Park and Rising Sea Level," memorandum to South Florida Water Management District, January 9, 1984.

18. P. J. Wilcoxen, "Coastal Erosion and Sea Level Rise: Implications for Ocean Beach and San Francisco's Westside Transport Project," *Coastal Zone Management Journal* 14 (1986):173–191.

CHAPTER 4

1. NEPA Sec. 102(2) (C), 42 U.S.C. §4332d (2) (C); 40 C.F.R. §1502.22, as amended; 51 C.F.R. §1562.5.

2. Earl Eiker, acting chief, Hydraulic and Hydrology Division, Directorate of Civil Works, U.S. Army Corps of Engineers, letter of March 21, 1986.

3. A. Alan Hill, chairman, Council on Environmental Quality, letter to Senator Albert Gore, March 3, 1987.

4. Ibid.

5. David Wirth, senior staff attorney, NRDC, letter to A. Alan Hill, chairman, Council on Environmental Quality, November 9, 1988.

6. P.L. 100–204, §§1101–1106, 101 Stat. 1407–09 (1987).

7. J. Hoffman, J. Wells, and J. Titus, *Projecting Future Sea Level Rise*, EPA 230-09-007 (Washington, D.C.: EPA, 1983), 6:41–43.

8. J. Titus et al., *Potential Impacts of Sea Level Rise on the Beach at Ocean City, Maryland*, EPA 230-10-85-013 (Washington, D.C.: EPA, 1985).

9. T. Kyper and R. Sorensen, "The Impact of Selected Sea Level Rise Scenarios on the Beach and Coastal Structures at Sea Bright, New Jersey," *Coastal Zone '85* 2 (1985):2645–2659.

10. T. Kana, B. Baca, and M. Williams, *Potential Impacts of Sea Level Rise on Wetlands Around Charleston, South Carolina*, EPA 230-10-85-014 (Washington, D.C.: EPA, 1984), 3:39.

11. R. Park, T. Armentano, and C. Cloonan, "Predicting the Effects of Sea Level Rise on Coastal Wetlands," in *Effects of Changes in Stratospheric Ozone and Global Climate*, vol. 4 (New York: UNEP/EPA, 1986).

12. J. Smith and D. Tirpak (eds.), *The Potential Effects of Global Climate Change on the United States*, Report to Congress (Washington, D.C.: EPA, p. 123).

13. 16 U.S.C. §1458(c) (1985).

14. 15 C.F.R. §923.102(a) (1).

15. 15 C.F.R. §923.102(b) (1), (2) (ii), (5) (i) and (ii).

16. 15 C.F.R. §923.103.

17. P. Klarin and M. Hershman, "Response of Coastal Zone Management Programs to Sea Level Rise in the United States," *Coastal Management* 18 (3) (Summer 1990) In press.

18. P.L. 90–448, 42 U.S.C. §4001–128(c) (1977 & Supp. 1988).

19. M. Simmons, "The Evolving National Flood Insurance Program," Congressional Research Report, 1988.

20. J. S. Bragg, "Foreword," in *A Unified National Program for Floodplain Management* (Washington, D.C.: Federal Emergency Management Agency, 1986), p. 1–4.

21. R. J. Burby and E. J. Kaiser, *An Assessment of Urban Floodplain Management in the United States: The Case for Land Acquisition in Comprehensive Floodplain Management*, Technical Report 1 (Association of State Floodplain Managers, 1987), p. 4.

22. Ibid., pp. 7, 9.

23. Eiker letter; see note 2.

CHAPTER 5

1. NRDC sent a questionnaire to the coastal managers of each of the coastal states in May 1986 requesting information concerning state and local plan-

ning efforts, including studies, plans, and legislation that states and local governments are undertaking to address problems ensuing from sea level rise. The extremely helpful responses by state officials form a significant part of the information presented in this section of the report.

2. P. Klarin and M. Hershman, "Response of Coastal Zone Management Programs to Sea Level Rise in the United States," *Coastal Management* 18 (3) (Summer 1990) In press.

3. Ibid.

4. This section is based on a study by the California Coastal Commission: *Draft Report: Planning for an Accelerated Sea Level Rise Along the California Coast*, prepared by Lesley C. Ewing, Jaime M. Michaels, and Richard J. McCarthy, June 26, 1989. This draft document was prepared by the staff of the California Coastal Commission to investigate possible effects on the California coast from an accelerated sea level rise. The report has not been approved by the commission.

5. This section is based on a study by P. Williams: *An Overview of the Impact of Accelerated Sea Level Rise on San Francisco Bay* (San Francisco: San Francisco Bay Conservation and Development Commission, 1985). The BCDC subsequently commissioned a follow-up study, funded by a federal Coastal Zone Management Act grant, of sea level rise in the Bay Area. The study, *Future Sea Level Rise: Predictions and Implications for San Francisco Bay* (1987), uses historical rates of sea level rise to project relative sea level rise at eleven locations in the San Francisco Bay area in twenty years (2007) and fifty years (2037). Even assuming only historical rates of sea level rise, the study projects increases in relative sea level rise by the year 2007 ranging from a low of 0.02 meter at Sausalito to a high of 0.6 meter at Alviso Slough. Projections for the year 2037 range from a low of 0.05 meter in Sausalito to 1.5 meters at Alviso Slough (p. 46). The study concludes, using projections of historical rates of sea level rise, that the Suisun Marsh managed wetlands will be threatened by saltwater intrusion; that the North Bay marshes may be submerged as a result of reduced sedimentation; that tidal marshes in the South Bay, north of the Dumbarton Bridge, are likely to suffer significant losses because of reduced sediment deposition and shoreline erosion; and that diked wetlands in the South Bay, below the Dumbarton Bridge, could revert to open water and mudflats if the area were returned to tidal action (pp. 6–8, 51–64).

6. This section is based on a study prepared by the Delaware River Basin Commission: C. Hull and J. Titus, *Greenhouse Effect, Sea Level Rise, and Salinity in the Delaware Estuary*, EPA 230-05-86-010 (Washington, D.C.: Environmental Protection Agency, 1986).

7. Ibid.

8. This section is based on the following study: *Effects on Hawaii of a Worldwide Rise in Sea Level Induced by the "Greenhouse Effect"* (Honolulu: Hawaii Coastal Zone Management Program, 1985).

9. This section is based on the following study: G. Geise, D. Aubrey, and P. Zeeb, *Passive Retreat of Massachusetts Coastal Upland Due to Relative Sea-Level Rise* (Massachusetts Coastal Zone Management, Massachusetts Coastal Submergence Program, 1987). See also G. Geise and D. Aubrey, "Losing Coastal Upland to Relative Sea-Level Rise: Three Scenarios for Massachusetts," *Oceanus* 30 (3) (1987):16–23.

10. N.Y. Environmental Conservation Law §8–0101 et seq. (West 1990).

11. N.Y. Environmental Conservation Law §8–0103 (West 1990).

12. Ibid., §113A-8. See also California Environmental Quality Act, Cal. Pub. Res. Code §§21100–21176 (West 1986 & Supp. 1990); Hawaii Environmental Impact Statements, Hawaii Rev. Stat. title 19 §343 (Hawaii 1985 & Supp. 1987); Massachusetts Environmental Policy Act, Mass. Gen. Laws Ann. ch. 30 §§61–62H (West 1979 & Supp. 1989); Washington State Environmental Policy Act, Wash. Rev. Code Ann. §43.21C.030 (West 1983 & Supp. 1989).

13. N.C. Gen. Stat. §113A–4(2) (1989) (published by Michie).

14. N.C. Gen. Stat. §113A–8 (1989).

15. 7 Del. Code Ann. §6801 (1988) (published by Michie); emphasis added.

16. Delaware Regulations Governing Beach Protection and the Use of Beaches §2.10.

17. Louisiana Coastal Resources Program, Final Environmental Impact Statement, Guideline 1.7(s), (t) (1980).

18. 15 N.C. Admin. Code 7H.0308(a) (1) (B).

19. Ibid., 7H.0308(a) (1) (A). North Carolina does allow beach bulldozing and placement of temporary sandbag structures to control erosion and also allows low-intensity offshore passive sand-trapping devices and beach nourishment for temporary erosion control (7H.0308(a) (2)–(5)). The regulations leave open the possibility that some other (new or innovative) form of erosion control might be permitted. Demonstration of "sound engineering for their planned purpose" and certification by a licensed engineer would be required for such structures. Any erosion control measure that interferes with public access to and use of ocean beaches or significantly increases erosion on adjacent properties is prohibited (7H.0308(a) (1) (D), (E)).

20. Code of Maine Rules, ch. 355 §3(F) (1988).

21. N.Y. Envtl. Conserv. Law §34–0102(2) (West 1984 & Supp. 1990).

22. Ibid., 23.22§34–0102(5).

23. N.Y. Envtl. Conserv. Law §34–0108(3) (a), (b), (c), (d) (McKinney 1984 & Supp. 1990).

24. 38 Me. Rev. Stat. Ann. §§480–C (West 1989).

25. Code of Maine Rules, ch. 355 §3 (preamble).

26. Ibid., §(3) (A) (2). Regardless of sea level rise, no new structure or addition to an existing structure may be located on or seaward of a frontal dune; §(B) (2).

27. Ibid., §3(B) (2) (C). Permit applicants will be required to submit geological surveys showing that the site will remain stable after allowing for a 0.9-meter rise in sea level over 100 years.

28. Fla. Stat. Ann. §§161 et seq. (West 1972 & Supp. 1990).

29. Ibid., §161.053(1).

30. Ibid., §161.053(2); see also §161.053(1) (purposes of the act). A similar case is South Carolina. The state does not have a setback requirement like Florida but instead uses a system of survey benchmarks. The benchmarks are established every 2,000 feet on all developed and recreational beaches for shoreline monitoring. The South Carolina Coastal Council's beach monitoring program provides that the state's beaches be surveyed every six months and following any storm. Long-term erosion due to sea level rise will be detected and assessed in this program.

31. Fla. Stat. Ann. §161.053 (West 1972 & Supp. 1989).

32. Fla. Stat. Ann. §161.053(6) (b) (West 1972 & Supp. 1990). Single-family dwellings are excepted as long as (1) the land was platted or subdivided by metes and bounds before October 1, 1985; (2) the owner does not own another adjacent landward parcel; (3) the home will be landward of the frontal dune structure; and (4) the home will be as far landward as possible without being seaward of or on the frontal dune. See Fla. Stat. Ann. §161.053(6) (c) (West 1972 & Supp. 1990).

33. Fla. Stat. Ann. §161.053(6) (b). The seasonal high-water line is defined as the line formed by the intersection of the rising shore and the elevation of 150 percent of the local mean tidal range above local mean high water; Fla. Admin. Code §16B–33.24(2) (h) (6).

34. Fla. Admin. Code §16B–33.24(1).

35. See R. Thomas, "Future Sea Level Rise and Its Early Detection by Satellite Remote Sensing," in *Effects of Changes in Stratospheric Ozone and Global Climate*, vol. 4 (New York: UNEP/EPA, 1986), p. 31.

36. Code of S. Car. §48–39–280 (Lawyers Cooperative Publishing Co. 1989 Supp.).

37. Code of S. Car. §48–39–290(B) (West 1976 & Supp. 1989).

38. P.L. 90–448, 42 U.S.C. §4001–128(c) (West 1977 & Supp. 1988).

39. For summaries of existing state floodplain management programs, see P. Bloomgren, *Strengthening State Floodplain Management: Appendix A to Volume 3* (Boulder: Natural Hazards Research and Applications Information Center, University of Colorado, 1982); J. Kusler, *Innovation in Local Floodplain Management: Appendix B to Volume 3* (Boulder: Natural Hazards Research and Applications Information Center, University of Colorado, 1982).

40. 38 Me. Rev. Stat. Ann. §§480 et seq. (West Supp. 190). Maine's Sand Dune Law is a provision of Maine's Coastal Wetlands Act. Since the Sand Dune Rules apply only within the sand dune system, the flood regulations described apply only within that system. Where rules apply because of flood

zones, the zones are based on FEMA flood insurance maps. The 100-year flood elevations must conform with the procedures established by FEMA in developing flood insurance rate maps (Code of Maine Rules, ch. 355 §1). Maine Geological Survey maps show both the sand dune system and flood zones. It should also be noted that the Sand Dune Rules do not necessarily differentiate between rules promulgated for flood protection and rules promulgated because of erosion concerns. However, all conditions of the rules must be satisfied. Thus a proposed structure that may be planned for a site that is not in a flood hazard area might not be safe from erosion and therefore could not be built at the proposed location.

41. Code of Maine Rules, ch. 355 §3.
42. Ibid., §1(AA).
43. Ibid., §3(B) (2) (a) (i).
44. Ibid., §1(A).
45. Ibid., §1(B).
46. Ibid., §3(B) (2) (b) (i), (ii).
47. Ibid., §3(B) (2) (d). The 4-foot elevation requirement is intended to ensure stability after allowing for a 0.9-meter sea level rise.
48. Ibid., §3(B) (2) (C).
49. New York's Floodplain Management Regulations, like those of other states, are based on FEMA 100-year flood maps; thus sea level rise is not taken into consideration in determining the 100-year flood.
50. N.Y. Admin. Code tit. 6 §500–502.9.
51. Ibid., §500.10(a) (3).
52. Ibid., §500.10(b) (2) and (3), (c).
53. Ibid., §500.10(d).
54. See §502.4 (regulations pertaining to state agencies).
55. Ibid., §500.10(a) (1) (ii).
56. Ibid., §500.10(a) (1) (iii), (b), (c), (d).
57. Ibid., §500.10(a) (2) (i) and (ii), (b), (c), (d).
58. Ibid., §502.4(a) (17) (i).
59. *Massachusetts Coastal Zone Management Program Coastal Policies: Coastal Hazards Policy* (8), vol. 1:119, 123 (1977); Mass. Admin. Code tit. 780 §744.0.
60. Cal. Pub. Res. Code §30240(b) (1987).
61. The manner in which the commission implements the guidelines—that is, as regulations or simply as guidelines—is currently the subject of litigation.
62. See California Coastal Commission Statewide Interpretive Guidelines, VII.
63. Md. Nat. Res. Code 14.15.09C(1).
64. Ibid., 14.15.09C(2).
65. Ibid., 14.15.09C(5) (d) (7). The "Critical Area" itself, in which activities are regulated according to the degree of development and environmental concerns, is defined to include "all land and water areas within 1,000 feet beyond the landward boundaries of State or private wetlands"; Md. Nat. Res. Code Ann. 8–1807(a) (2). See also 14.15.05A(2) (defining types of

development areas) and 14.15.05C(5) (defining policies that must be followed in developing Critical Area programs). Other states have also established wetlands buffers; see Appendix E.

66. See generally N.Y. Admin. Code tit. 6 §661.5(6). Intertidal marshes and coastal freshwater marsh wetlands are the "most biologically productive [and] their intertidal location also makes them among the most effective wetland zones for flood and hurricane and storm protection"; N.Y. Admin. Code tit. 6 §661.2(d). Thus only limited types of land use and development are compatible with these areas. High marsh and salt meadow tidal wetlands are similarly ranked. Coastal shoals, bars and flats, and littoral zones are classified as more variable and "more extensive and intensive uses" may be appropriate; N.Y. Admin. Code tit. 6 §661.2(e).

67. N.Y. Admin. Code tit. 6 §661.4(b) (1).

68. The regulations state that "a wide variety of uses may be compatible with these areas, provided such uses do not adversely affect adjacent and nearby tidal wetlands"; N.Y. Admin. Code tit. 6 §661.2(j). Some uses that would clearly be incompatible with wetlands migration and are considered incompatible in existing wetlands are considered only "presumptively incompatible" in adjacent areas. These uses include storage of any chemical, petrochemical, or other toxic material, including any pesticide, and disposal of solid wastes; N.Y. Admin. Code tit. 6 §661.5(b) (54), (56).

69. N.Y. Admin. Code tit. 6 §661.2(j).

70. N.J. Admin. Code §7:7E–3.26(a).

71. N.J. Admin. Code §7:7E–3.26(b).

72. Ibid. The current buffer requirements were designed to protect wetlands as they currently exist. Thus the rationale for the buffer requirement states that "development adjacent to wetlands can adversely affect the wetlands through increased runoff, sedimentation, and introduction of pollutants" (N.J. Admin. Code 7:7E–3.26(c)). These effects on wetlands are taken into consideration in determining whether a proposed development in the buffer will have a "significant adverse effect." To protect the wetlands from sea level rise, the effect of development in the buffer zone on wetlands migration would also have to be considered. It should be noted, too, that the Wetlands Act itself does not confer jurisdiction beyond the wetlands boundary and that the jurisdiction for the buffer is the Coastal Area Facility Review Act (CAFRA), rather than the Wetlands Act. Thus the buffer does not apply to those facilities exempted from CAFRA, notably buildings of twenty-four or fewer dwelling units. See N.J. Admin. Code tit. 7 §7–2.1(a) (1984). Obviously this exemption is a major limitation on the efficacy of the buffer in allowing wetlands migration.

73. California Water Code §§30000–33901 (West 1984 & Supp. 1987).

74. Ibid., §31020.

75. Ibid., §31021.

76. Fla. Stat. Ann. §§373.012–373.619 (West 1974 & Supp. 1987).

77. Ibid., §373.036(1).

78. Ibid.

79. Florida's Water Resources Act authorizes establishment of five water management districts; Fla. Stat. Ann. §373 (West 1985 & Supp. 1987).

80. Ibid., §373.033(1).

81. Ibid., §373.59.

82. Ibid., §373.59(3). Pennsylvania's Open Space Lands Act authorizes the Department of Forests and Waters or any county to acquire any interest in real property for the purpose of, among other objectives, protecting and conserving water resources and watersheds.

83. Ga. Code Ann. §12–5–95 (1982).

84. Ibid., §12–5–96(a).

85. N.J. Stat. Ann. §13:1E–49 et seq.

86. N.J. Stat. Ann. §13:1E–57(a) (2)–(4).

87. N.J. Stat. Ann. §13:1E–57(a).

88. Private facilities are regulated by permit. See Conn. Gen. Stat. Ann. §§22a–117 (West 1985 & Supp. 1987).

89. Conn. Gen. Stat. Ann. §22a–134cc(3) (West Supp. 1987).

90. Ibid., §22a–134cc(3)(C) and (E).

91. Code of Maine Rules, ch. 854 §6–096. See also Md. Nat. Res. Code Ann. §3–705(b) (1) (1983) (requiring consideration of environmental factors as they apply to a proposed site in deciding whether to issue a certificate of public necessity for establishment of a hazardous waste facility in Maryland); La. Rev. Stat. Ann. §1153.2(A) (instructing Louisiana's Hazardous Waste Advisory Board to assess the impact of a proposed hazardous waste facility on the environment and to determine whether a site for a proposed facility is environmentally suitable).

92. U.S. Environmental Protection Agency and Louisiana Geological Survey, *Saving Louisiana's Coastal Wetlands: The Need for a Long-Term Plan of Action*, EPA–230–02–87–026 (Washington, D.C.: EPA, 1987), p. 9. The Mississippi River deltaic plain lost 29 square miles per year between 1955 and 1978. The rate of wetland loss has been increasing geometrically over the last century. The rate of wetland loss in 1985 for the deltaic plain alone was 45 square miles per year. The Chenier Plain in southwestern Louisiana is losing an additional 10 square miles per year, bringing the total land loss within the coastal zone to approximately 55 square miles per year.

93. A. Sallenger, Jr., "Louisiana Barrier Island Erosion Study," *Coastal Sediments '87* (New York: American Society of Civil Engineers, 1987), p. 1503.

94. J. van Beck and K. M. Arendt, *Louisiana's Eroding Coastline: Recommendations for Protection* (Baton Rouge: Coastal Management Section, Louisiana Department of Natural Resources, 1982), pp. 12–13; Coalition to Restore Coastal Louisiana, *Coastal Louisiana: Here Today and Gone Tomorrow?* (draft for public review, April 1987), p. 9; EPA, *Saving Louisiana's Coastal Wetlands*, pp. 16–18.

95. EPA, *Saving Louisiana's Coastal Wetlands*, p. 31.
96. van Beck and Arendt, *Louisiana's Eroding Coastline*, p. 3; CRCL, *Here Today and Gone Tomorrow?*, pp. 9–10; EPA, *Saving Louisiana's Coastal Wetlands*, p. 20.
97. EPA, *Saving Louisiana's Coastal Wetlands*, p. 22.
98. Ibid., p. 22. Estuaries are similarly threatened by wave-generated erosion (p. 25). Boat wakes and the ebb and flow of the tides within canals also contribute to erosion along the canal banks.
99. S. Penland and J. Suter, "The Erosion and Protection of Louisiana's Barrier Islands," in S. Penland and J. Suter (eds.), *Barrier Shoreline Geology, Erosion and Protection in Louisiana, Coastal Sediments '87* (New York: American Society of Civil Engineers, 1987), pp. 1–4. Since the end of the Holocene transgressions, the Mississippi River has built its plain by a process of delta building followed by abandonment and barrier shoreline generation. The delta is built until eventually decreasing river gradients make the river's distributing course hydraulically inefficient, resulting in abandonment in favor of a more efficient course. The abandoned delta then subsides, saltwater intrudes, displacing freshwater environments landward, and coastal processes replace deltaic processes. These coastal processes rework the seaward margin of the abandoned delta, generating a sandy barrier island.
100. Penland and Suter, "Erosion and Protection." The formation of the barrier arc is actually the second stage in a three-stage model for the development of transgressive Mississippi River delta barrier shorelines proposed by Penland and Boyd. See S. Penland and R. Boyd, "Assessment of Geological and Human Factors Responsible for Louisiana Coastal Barrier Erosion," in *Proceedings of the Conference on Coastal Erosion and Wetland Modification in Louisiana: Causes, Consequences and Options* (Washington, D.C.: U.S. Department of Interior Fish and Wildlife Service, 1982), pp. 20–59; S. Penland et al., "Deltaic Barrier Development on the Louisiana Coast," in *Transactions of the Gulf Coast Association of Geological Societies* 31 (1981):471–476. In the first stage, compactional subsidence generates a relative sea level rise and coastal processes transform the once active delta into an erosional headland with flanking barrier islands. Barrier islands are built through a spit-breaching process. See Penland and Suter, "Erosion and Protection," pp. 1–4.
101. Penland and Suter, "Erosion and Protection," pp. 1–4.
102. CRCL, *Here Today and Gone Tomorrow?*, p. 38.
103. Sallenger et al., "Louisiana Barrier Island Erosion Study."
104. Penland and Suter, "Erosion and Protection," pp. 1–3.
105. EPA, *Saving Louisiana's Coastal Wetlands*, p. 36 (citing Hoffman et al., *Projecting Future Rise*, high and low estimates). Relative sea level rise in Louisiana by the year 2050 is projected to be from 0.9 to 1.25 meters (approximately 3 to 4 feet).

106. 1981 La. Acts 41 (special session).
107. Response of David Chambers, chief, Coastal Protection Section, Louisiana Department of Natural Resources, to NRDC's questionnaire titled "State Response to Sea Level Rise" (hereafter Chambers, "Response to NRDC Questionnaire"). See EPA, *Saving Louisiana's Coastal Wetlands*, pp. 62–76.
108. A parish is an administrative subdivision in Louisiana that corresponds to a county in other states.
109. Chambers, "Response to NRDC Questionnaire"; see also EPA, *Saving Louisiana's Coastal Wetlands*, pp. 62–65.
110. Chambers, "Response to NRDC Questionnaire"; see also EPA, *Saving Louisiana's Coastal Wetlands*, pp. 65, 73. A project is also planned to repair damages to Timbalier Island (Terrebonne Parish) caused by Hurricane Juan in 1985. The state will provide funds to replace beach materials that were eroded from the east and west ends of the present Timbalier Island seawall. Additional state and private funds may be used to fill adjacent canals and slips in order to reduce the effects of future storms on the island. At Grand Isle (Jefferson Parish), surveys are under way to determine the extent of the damage done to the Grand Isle sand dune during the 1985 hurricane season. A plan that will involve replacement of eroded dune segments, extension of the jetties on the east and west ends of the island, and installation of breakwater structures in critical areas is presently being developed. Other barrier island and shoreline restoration projects, contingent on funding availability, have also been planned. These include restoration of the Plaquemines Parish barrier shoreline, restoration of Timbalier and East Timbalier, and restoration of the shoreline between Belle Pass and Caminda Pass in LaFourche and Jefferson parishes.
111. Chambers, "Response to NRDC Questionnaire"; EPA, *Saving Louisiana's Coastal Wetlands*, p. 69.
112. EPA, *Saving Louisiana's Coastal Wetlands*, p. 69.
113. Ibid. Several freshwater diversion projects, which will introduce fresh water and sediments into deteriorating marshes, are also planned. The Caervon Freshwater Diversion (St. Bernard Parish), a joint state/federal effort designed to reduce marsh loss by approximately 16,000 acres over the next fifty years, was authorized by Congress in 1985. The project is an attempt to introduce fresh water and sediments into the marshes and estuarine waters of the Breton Sound basin. Advanced engineering and design have been completed and the project is now ready for construction. The Pass a Loutre Marsh creation project (Plaquemines Parish) entails artificially breaching the natural levees of the Mississippi River and its distributaries near the mouth of the river to allow fresh water and sediments to reach the areas outside the levees, filling open bays and ponds, and creating marshlands. Three diversions have been created within the Pass a Loutre Wildlife Management Area and are presently allowing

freshwater and sediment input into shallow subsiding bays. See Chambers, "Response to NRDC Questionnaire"; EPA, *Saving Louisiana's Coastal Wetlands*, pp. 68–69.

114. See, for example, F. Quinn, "Likely Effects of Climate Changes on Water Levels in the Great Lakes," in *Preparing for Climate Change—Proceedings of the First North American Conference on Preparing for Climate Change: A Cooperative Approach* (Washington, D.C.: Climate Institute, 1987), p. 481; S. Cohen, "The Effects of Climate Changes on the Great Lakes," in *Effects of Changes in Stratospheric Ozone and Global Climate*, vol. 3 (New York: UNEP/EPA, 1986), p. 163; J. Bruce, "Great Lakes Levels and Flows: Past and Future," *Journal of Great Lakes Research* 10 (1984):126.

115. Quinn, "Great Lakes," p. 481.

116. Ibid. See also S. Cohen, "How Climate Change in the Great Lakes Region May Affect Energy, Hydrology, Shipping and Recreation," in *Preparing for Climate Change*, pp. 460, 466.

117. Quinn, "Great Lakes," pp. 481–482. Precipitation over the Great Lakes is actually expected to increase. The increase in precipitation, however, will not be sufficient to offset losses from evapotranspiration on land surfaces and evaporation from the lakes' surfaces; see Cohen, "How Climate Change in the Great Lakes Region," p. 466.

118. Cohen, "The Effects of Climate Changes," p. 168. This scenario is based on the Goddard Institute for Space Studies (GISS) model of carbon-dioxide-induced global warming.

119. Ibid. This scenario is based on a model developed by the Geophysical Fluid Dynamics Lab (GFDL). It has also been estimated that a 3°C temperature rise coupled with a 6.5 percent increase in precipitation would result in a decrease in the net basin supply to Lake Superior of 10 percent. A 3°C temperature rise with no precipitation increase would reduce the Lake Erie net basin supply by 33 percent. See F. Quinn and T. Croley II, "Climate Basin Water Balance Models for Great Lakes Forecasting and Simulation," in *Proceedings of the Fifth Conference on Hydrometeorology* (American Meteorological Society, 1983), pp. 218–228.

120. Quinn, "Great Lakes," p. 5. Navigation may, however, be aided by a shorter ice season, which could lead to an extension of the current navigation season. Such an effect would contribute to better vessel utilization and less stockpiling.

121. Quinn, "Great Lakes," p. 5. It has also been projected, however, that global warming will result in a reduction in energy demand that would actually exceed the losses in hydroelectric power production; see Cohen, "The Effects of Climate Changes," p. 163.

122. Quinn, "Great Lakes," p. 485.

123. Cohen, "The Effects of Climate Changes," pp. 173–179.

124. J. Raoul and Z. Goodwin, "Climate Changes—Impacts in Great Lakes Levels and Navigation," in *Preparing for Climate Change*, pp. 488, 497. See

also Cohen, "How Climate Change in the Great Lakes Region," pp. 467–468.

125. Quinn, "Great Lakes," pp. 5–6.

CHAPTER 6

1. Cairo Compact and Panel Reports, from the World Conference on Preparing for Climate Change, December 17–21, 1989, convened by the Climate Institute, United Nations Environment Program, and the government of Egypt; "The Noordwijk Declaration on Atmospheric Pollution and Climatic Change of November 1989," *U.S. IPCC News* 6 (December 1989); Chairman's summary, Tokyo Conference on the Global Environment and Human Response Toward Sustainable Development (September 11–13, 1989); Conference statement, International Conference on "Global Warming and Climate Change: Perspectives from Developing Countries," organized by Tata Energy Research Institute and the Woods Hole Research Center, cosponsored by United Nations Environment Program and World Resource Institute (New Delhi, February 21–23, 1989); Conference statement, The Changing Atmosphere, Toronto, June 1988; IPCC Workshop, Report of the Coastal Resource Management Subgroup on Adaptive Options and Policy Implications of Sea Level Rise and Other Coastal Impacts of Global Climate Change (Miami, November 27–December 1, 1989).

2. World Meteorological Organization/United Nations Environment Program: Intergovernmental Panel on Climate Change, *Formulation of Response Strategies*. Prepared for IPCC by Working Group III (Washington, D.C.: Island Press, 1991).

3. Ibid.

4. *Quarterly Economic Review of Egypt* (London: Economist Publications, 1985), annual supplement.

5. J. Broadus et al., "Rising Sea Level and Damming of Rivers: Possible Effects in Egypt and Bangladesh, in *Effects of Changes in Stratospheric Ozone and Global Climate*, vol. 4 (New York: UNEP/EPA, 1986), pp. 165–190.

6. E. G. Jansen, *Rural Bangladesh: Competition for Scarce Resources* (Bergen, Norway: Michelsen Institute, 1983).

7. Broadus et al., "Egypt and Bangladesh," p. 171.

8. Ibid., pp. 177–179.

9. Ibid., p. 179.

10. T. Goemans, "The Sea Also Rises: The Ongoing Dialogue of the Dutch with the Sea," in *Effects of Changes in Stratospheric Ozone and Global Climate*, vol. 4 (New York: UNEP/EPA, 1986), pp. 47–56.

11. This example comes from Goemans, "The Sea Also Rises: The Ongoing Dialogue of the Dutch with the Sea," in *Effects of Changes in Stratospheric Ozone and Global Climate*, vol. 4 (New York: UNEP/EPA, 1986), pp. 47–56.

BIBLIOGRAPHY

Abrahamson, D. (ed.). *The Challenge of Global Warming*. Washington, D.C.: Island Press, 1989.

"Assessment of Geological and Human Factors Responsible for Louisiana Coastal Barrier Erosion." In *Proceedings of the Conference on Coastal Erosion and Wetland Modification in Louisiana: Causes, Consequences and Options*. Washington, D.C.: U.S. Department of Interior Fish and Wildlife Service, 1982.

Barth, M., and J. Titus. *Greenhouse Effect and Sea Level Rise*. New York: Van Nostrand Reinhold, 1984.

Berger, A. "Support for the Astronomical Theory of Climate Change." *Nature* 268 (1977):44–45.

Broadus, J., et al. "Rising Sea Level and Damming of Rivers: Possible Effects in Egypt and Bangladesh." In *Effects of Changes in Stratospheric Ozone and Global Climate*, vol. 4. New York: U.N. Environment Program/U.S. Environmental Protection Agency, 1986.

Broecker, W. "Unpleasant Surprises in the Greenhouse?" *Nature* 328 (1987):123–126.

Broecker, W., et al. "Fate of Fossil Fuel Carbon Dioxide and the Global Carbon Budget." *Science* 206 (1979):409–418.

Brown, D. "Factoring Sea Level Rise into Coastal Zone Management." In *Preparing for Climate Change*. Washington, D.C.: Climate Institute, 1987.

Bruce, J. "Great Lakes Levels and Flows: Past and Future." *Journal of Great Lakes Research* 10 (1984):126.

Bruun, P. "Sea Level Rise as a Cause of Shore Erosion." *Proceedings of the American Society of Engineers and Journal Waterways Harbors Division* 88 (1962):117–130.

Bruun, P. "Worldwide Impacts of Sea Level Rise on Shorelines." In *Effects of Changes in Stratospheric Ozone and Global Climate*, vol. 4. New York: U.N. Environment Program/U.S. Environmental Protection Agency, 1986.

Burby, R., and E. Kaiser. *An Assessment of Urban Floodplain Management in the United States: The Case for Land Acquisition in Comprehensive Floodplain Management*. Technical Report 1. Association of State Floodplain Managers, 1987.

Caldwell, C., and R. C. Cavanagh. *The Decline of Conservation at California Utilities: Causes, Costs and Remedies*. Natural Resources Defense Council, 1989.

Cavanagh, R., D. Goldstein, and R. Watson. *One Last Chance for National Energy Policy*. Washington, D.C.: Natural Resources Defense Council, 1988.

Cavanagh, R., C. Caldwell, D. Goldstein, and R. Watson. "National Energy Policy: From Costly Chaos to Least-Cost Planning." *World Policy Journal* (1989):239–264.

Climate Institute. *Climate Change Priorities*. Report of the Committee on Climate Funding Priorities. Washington, D.C.: Climate Institute, 1989.

Cohen, S. "The Effects of Climate Changes on the Great Lakes." In *Effects of Changes in Stratospheric Ozone and Global Climate*, vol. 3. New York: U.N. Environment Program/U.S. Environmental Protection Agency, 1986.

"Cooling the Greenhouse: Vital First Steps to Combat Global Warming." Washington, D.C.: Natural Resources Defense Council, Spring, 1989.

Crutzen, P., and L. Gidel. "A Two-Dimensional Photochemical Model of the Atmosphere 2: The Tropospheric Budget of the Anthropogenic Chlorocarbons, CO, CH_4, CHCl and the Effect of Various NO_x Sources on Tropospheric Ozone." *Journal of Geophysical Research* 88 (1983):6641–6661.

Cunnold, D., et al. "The Atmospheric Lifetime Experiments 3 and 4: Lifetime Methodology and Application to Three Years of CFC Data." *Journal of Geophysical Research* 88 (1983):8379–8414.

Delmas, R., J. Ascencio, and M. Legrand. "Polar Ice Evidence That Atmospheric CO_2 20,000 BP Was 50% of Present." *Nature* 284 (1980):155–157.

deSylva, D. "Increased Storms and Estuarine Salinity and Other Ecological Impacts of the Greenhouse Effect." In *Effects of Changes in Stratospheric Ozone and Global Climate*, vol. 4. New York: U.N. Environment Program/U.S. Environmental Protection Agency, 1986.

Dickinson, R. "How Will Climate Change?" In B. Bolin et al. (eds.), *The Greenhouse Effect, Climate Change, and Ecosystems*. New York: Wiley, 1986.

Dickinson, R., and R. Cicerone. "Future Global Warming from Atmospheric Trace Gases." *Nature* 319 (1986):109–115.

Dolan, R., and H. Lins. "Beaches and Barrier Islands." *Scientific American* 257 (1987):68–77.

Doniger, D. *Saving the Ozone Layer: A Citizen Action Guide*. Washington, D.C.: Natural Resources Defense Council, 1988.

Doniger, D., and D. Wirth. "Cooling the Chemical Summer." In *Effects of Changes in Stratospheric Ozone and Global Climate*, vol. 1. New York: U.N. Environment Program/U.S. Environmental Protection Agency, 1986.

Donovan, D., and E. Jones. "Causes of Worldwide Changes in Sea Level." *Geological Society of London Journal* 136 (1979):187–192.

Douglas, P., and R. H. Stroud. *A Symposium on the Biological Significance of Estuaries*. Washington, D.C.: Sport Fishing Institute, 1971.

Effects on Hawaii of a Worldwide Rise in Sea Level Induced by the "Greenhouse Effect." Honolulu: Hawaii Coastal Zone Management Program, 1985.

Emanuel, K. "The Dependence of Hurricane Intensity on Climate." *Nature* 326 (1987):483–485.

Environment Canada. *Conference Proceedings: The Changing Atmosphere Conference*. Geneva: World Meteorological Organization, 1989.

Geise, G., D. Aubrey, and P. Zeeb. *Passive Retreat of Massachusetts Coastal Upland Due to Relative Sea-Level Rise*. Massachusetts Coastal Zone Management, Massachusetts Coastal Submergence Program, 1987.

Geise, G., et al. "Losing Coastal Upland to Relative Sea Level Rise: Three Scenarios for Massachusetts." *Oceanus* 30 (1987):16–23.

Gibney, F. "Louisiana's Bayou Blues." *Newsweek*, June 22, 1987.

Glynn, P. "Widespread Coral Mortality and the 1982–83 El Niño Warming Event." *Environmental Conservation* 11 (2) (1988):133–146.

Goemans, T. "The Sea Also Rises: The Ongoing Dialogue of the Dutch with the Sea." In *Effects of Changes in Stratospheric Ozone and Global Climate*, vol. 4. New York: U.N. Environment Program/U.S. Environmental Protection Agency, 1986.

Gornitz, V., and P. Kaneiruk. "Assessment of Global Coastal Hazards from Sea Level Rise." In *Proceedings of Sixth Symposium on Coastal and Ocean Management*. ASCE, 1989.

Gornitz, V., S. Lebedeff, and J. Hansen. "Global Sea Level Trend in the Past Century." *Science* 215 (1982):1611–1614.

Hansen, J. "Prediction of Near-Term Climate Evolution: What Can We Tell Decision-Makers Now?" Paper presented at the First North American Conference on Preparing for Climate Change: A Cooperative Approach, Washington, D.C., 1987.

Hansen, J., and S. Lebedeff. *Journal of Geophysical Research* 92 (1987):13,345–13,372.

Hansen, J., W. Wang, and A. Lacis. "Mt. Agung Eruption Provides Test of Global Climate Perturbation." *Science* 199 (1987):1065–1068.

Hansen, J., J. Wells, and J. Titus. "Future Global Warming and Sea Level Rise." In G. Sigbjarnarson (ed.), *Iceland Coastal and River Symposium Proceedings*. Reykjavik: National Energy Authority, 1986.

Hansen, J., et al. "Climate Sensitivity to Increasing Greenhouse Gases." In M. C. Barth and J. G. Titus (eds.), *Greenhouse Effect and Sea Level Rise: A Challenge for This Generation*. New York: Van Nostrand Reinhold, 1984.

Hansen, J., et al. "The Greenhouse Effect: Projections of Global Climate Change." In *Effects of Changes in Stratospheric Ozone and Global Climate*, vol. 1. New York: U.N. Environment Program/U.S. Environmental Protection Agency, 1986.

Hoffman, J., J. Wells, and J. Titus. *Projecting Future Sea Level Rise*. Washington, D.C.: U.S. Environmental Protection Agency, 1983.

Holeman, J. "The Sediment Yield of Major Rivers of the World." *Water Resources Research* 4 (1968):737–747.

Hull, C., H. Thatcher, and R. C. Tortoriello. "Salinity in the Delaware Estuary." In *Greenhouse Effect, Sea Level Rise, and Salinity in the Delaware Estuary*. Washington, D.C.: U.S. Environmental Protection Agency and Delaware River Basin Commission, 1986.

Hyman, W. *Potential Impacts of the Greenhouse Effect on the Infrastructure of Greater Miami, Florida*. Washington, D.C.: Urban Institute, 1988.

Jansen, E. *Rural Bangladesh: Competition for Scarce Resources.* Bergen, Norway: Michelsen Institute, 1983.

Jones, P., T. Wigley, and P. Wright. "Global Temperature Variations Between 1861 and 1984." *Nature* 3 (1986):430–434.

Kana, T., B. Baca, and M. Williams. *Potential Impacts of Sea Level Rise on Wetlands Around Charleston, South Carolina.* Washington, D.C.: U.S. Environmental Protection Agency, 1984.

Kana, T., et al. "The Physical Impact of Sea Level Rise in the Area of Charleston, South Carolina." In M. C. Barth and J. G. Titus (eds.), *Greenhouse Effect and Sea Level Rise: A Challenge for This Generation.* New York: Van Nostrand Reinhold, 1984.

Kerr, R. "Is the Greenhouse Here?" *Science* 239 (1988):559.

Klarin, P., and M. Hershman. "Response of Coastal Zone Management Programs to Sea Level Rise in the United States," *Coastal Management* 18 (3) (Summer 1990).

Knox, F., and M. McElroy. "Changes in Atmospheric CO_2: Influence of the Marine Biota at High Latitude." *Journal of Geophysical Research* 89 (1984):4629–4637.

Ko, M., and N. Sze. "A 2-D Model Calculation of Atmospheric Lifetimes for N_2O, CFC-11, CFC-12." *Nature* 287 (1982):317–319.

Kyper, T., and R. Sorensen. "The Impact of Selected Sea Level Rise Scenarios on the Beach and Coastal Structures at Sea Bright, New Jersey." *Coastal Zone '85* 2 (1985):2645–2659.

Lachenbruch, A., and B. Marshall. "Changing Climate: Geothermal Evidence from Permafrost in the Alaskan Arctic." *Science* 234 (1986):689–696.

Lane, P., and Associates. "Preliminary Study of the Possible Impacts of a One Metre Rise in Sea Level at Charlottetown, Prince Edward Island." Report prepared under contract for Environment Canada, 1986.

Lawrence, B. "Towards a National Coastal Policy." *Environmental Law Reporter* (October 1987):10404.

Leatherman, S. "Barrier Island Evolution in Response to Sea Rise: A Discussion." *Journal of Sedimentary Petrology* 53 (1983):1026–1031.

Lorius, C., N. Barkov, J. Jouzel, Y. Korotkevich, V. Kotlyakov, and D. Raynaud. "Antarctic Ice Core: CO_2 and Climatic Change Over the Last Climate Cycle." *EOS* 69 (1988):681–684.

Manzer, M. "An Overview of the Potential Effects of Sea Level Rise on Coastal Communities and Infrastructure in Atlantic Canada." Discovery Consultants Ltd., 1988.

Martec Limited. "Effects of a One Metre Rise in Mean Sea Level at Saint John, New Brunswick, and the Lower Reaches of the Saint John River." Report prepared under contract for Environment Canada, 1987.

Meier, M. "News and Views: Reduced Rise in Sea Level." *Nature* 343 (1990):115–116.

Mintzer, I. *A Matter of Degrees: The Potential for Controlling the Greenhouse Effect.* Washington, D.C.: World Resources Institute, 1987.

Moffat and Nichol. *Future Sea Level Rise: Predictions and Implications for San Francisco Bay.* San Francisco: San Francisco Bay Conservation and Development Commission, 1987.

Molina, M., and F. Rowland. "Stratospheric Sink for Chlorofluoromethanes: Chlorine Catalyzed Destruction of Ozone." *Nature* 249 (1974):810–812.

National Academy of Sciences. *Changing Climate.* Washington, D.C.: National Academy Press, 1983.

National Council on Public Works Improvement. *Fragile Foundations: A Report on America's Public Works.* Final report to the president and Congress. Washington, D.C.: National Council on Public Works Improvement, 1988.

National Research Council. *Carbon Dioxide and Climate: A Scientific Assessment.* Washington, D.C.: National Research Council, 1979.

National Research Council. *Responding to Changes in Sea Level: Engineering Implications.* Washington, D.C.: National Academy of Sciences Press, 1987.

Natural Resources Defense Council. *Cooling the Greenhouse: Vital First Steps to Combat Global Warming.* Washington, D.C.: Natural Resources Defense Council, 1989.

Newall, R., and A. Deepak. *Mt. St. Helens Eruptions of 1980: Atmospheric Effects and Potential Climate Impact.* Washington, D.C.: NASA Scientific and Technical Information Branch, 1982.

Odum, E. *Fundamentals of Ecology.* Philadelphia: Saunders, 1971.

Oerlemans, J. "A Projection of Future Sea Level." *Climatic Change* 15 (1989):151–174.

Park, R., T. Armentano, and C. Cloonan. "Predicting the Effects of Sea Level Rise on Coastal Wetlands." In *Effects of Changes in Stratospheric Ozone and Global Climate,* vol. 4. New York: U.N. Environment Program/U.S. Environmental Protection Agency, 1986.

Penland, S., and J. Suter. "The Erosion and Protection of Louisiana's Barrier Islands." In *Barrier Shoreline Geology, Erosion and Protection in Louisiana, Coastal Sediments '87.* New York: American Society of Civil Engineers, 1987.

Penland, S., et al. "Deltaic Barrier Development on the Louisiana Coast." In *Transactions of the Gulf Coast Association of Geological Societies* 31 (1981):471–476.

Pilkey, O., Jr., and J. Howard. "Saving the American Beach: A Position Paper by Concerned Coastal Geologists." Paper presented at the Skidaway Institute Conference on America's Eroding Shorelines, 1981.

Pittman III, W. "Relationship Between Eustasy and Stratigraphic Sequences of Passive Margins." *Geological Society of America Bulletin* 89 (1978):1389–1403.

Platt, R. "Metropolitan Flood Loss Reduction Through Regional Special Districts." *Journal of the American Planning Association* 52 (1979):467–479.

Preventing Coastal Flood Disasters: The Role of the States and Federal Response. Special Publication 7. Washington, D.C.: Natural Hazards Research and Applications Information Center, 1983.

Prickett, G., and D. Wirth. *Banking on Sustainable Energy: The Multilateral Devel-*

opment Banks, Energy Efficiency, and Renewable Energy. Washington, D.C.: Natural Resources Defense Council, Inc., 1990.

Quinn, F. "Likely Effects of Climate Changes on Water Levels in the Great Lakes." In *Preparing for Climate Change—Proceedings of the First North American Conference on Preparing for Climate Change.* Washington, D.C.: Climate Institute, 1987.

Ramanathan, V., et al. "Trace Gas Trends and Their Potential Role in Climate Change." *Journal of Geophysical Research* 90 (1985):5547–5567.

Revelle, R. "Probable Changes in Sea Level Resulting from Increased Atmospheric Carbon Dioxide." In *Changing Climate.* Washington, D.C.: National Academy Press, 1983.

Sallenger, A., Jr. "Louisiana Barrier Island Erosion Study." In *Coastal Sediments '87.* New York: American Society of Civil Engineers, 1987.

Shackleton, N., et al. "Carbon Isotope Data in Core V19–30 Confirm Reduced Carbon Dioxide Concentration in Ice Age Atmosphere." *Nature* 306 (1983):319–322.

Sheiman, D., D. Doniger, and L. Dator. *A Who's Who of American Ozone Depleters: A Guide to 3,014 Factories Emitting Three Ozone-Depleting Chemicals.* Washington, D.C.: Natural Resources Defense Council, 1990.

Siegenthaler, U., and T. Wenk. "Rapid Atmospheric CO_2 Variations and Ocean Circulation." *Nature* 308 (1984):624–626.

Sorensen, R., et al. "Control of Erosion, Inundation, and Salinity Intrusion Caused by Sea Level Rise: A Challenge for This Generation." In M. C. Barth and J. G. Titus (eds.), *Greenhouse Effect and Sea Level Rise: A Challenge for This Generation.* New York: Van Nostrand Reinhold, 1984.

Stokoe, P. "Socio-Economic Assessment of the Physical and Ecological Impacts of Climate Change on the Marine Environment of the Atlantic Region of Canada: Phase 1." Report prepared under contract for Environment Canada, 1987.

Thomas, R. "Future Sea Level Rise and Its Early Detection by Satellite Remote Sensing." In *Effects of Changes in Stratospheric Ozone and Global Climate,* vol. 4. New York: U.N. Environment Program/U.S. Environmental Protection Agency, 1986.

Titus, J., et al. *Potential Impacts of Sea Level Rise on the Beach at Ocean City, Maryland.* Washington, D.C.: U.S. Environmental Protection Agency, 1985.

Topping, J. "Climate Change and Stratospheric Ozone Depletion: Need for More Than the Current Minimalist Response." In *Effects of Changes in Stratospheric Ozone and Global Climate,* vol. 1. New York: U.N. Environment Program/U.S. Environmental Protection Agency, 1986.

U.S. Environmental Protection Agency. "State Wetland Protection Programs—Status and Recommendations." Washington, D.C.: EPA, 1986.

U.S. Environmental Protection Agency. "The Potential Effects of Global Climate Change on the United States." Draft report. Washington, D.C.: EPA, 1988.

U.S. Environmental Protection Agency and Louisiana Geological Survey. "Saving Louisiana's Coastal Wetlands: The Need for a Long-Term Plan of Action." Washington, D.C.: EPA, 1987.

U.S. Water Resources Council. *National Report*. Washington, D.C.: Water Resources Council, 1978.

van Beck, J., and K. M. Arendt. *Louisiana's Eroding Coastline: Recommendation for Protection*. Baton Rouge: Louisiana Department of Natural Resources, Coastal Management Section, 1982.

Vellinga, P., and S. Leatherman. "Sea Level Rise, Consequences and Policies." *Climatic Change* 15 (1989):175–189.

Ward, J., R. Hardt, and T. Kuhnle. "Farming in the Greenhouse." *What Global Warming Means for American Agriculture*. Washington, D.C.: Natural Resources Defense Council, 1989.

Watson, R. *Oil and Conservation Fact Sheet: A Least-Cost Planning Perspective*. Washington, D.C.: Natural Resources Defense Council, 1988.

Watts, A. "Tectonic Subsidence, Flexure and Global Changes of Sea Level." *Nature* 297 (1982):469–474.

Wigley, T., P. Jones, and P. Kelly. "Empirical Climate Studies: Warm World Scenarios and the Detection of Climatic Change Induced by Radioactively Active Gases." In B. Bolin et al. (eds.), *The Greenhouse Effect, Climatic Change, and Ecosystems*. New York: Wiley, 1986.

Wilcoxen, P. "Coastal Erosion and Sea Level Rise: Implications for Ocean Beach and San Francisco's Westside Transport Project." *Coastal Zone Management Journal* 14 (1986):173–191.

Williams, P. *An Overview of the Impact of Accelerated Sea Level Rise on San Francisco Bay*. San Francisco: San Francisco Bay Conservation and Development Commission, 1985.

Wirth, D. "Climate Chaos." *Foreign Policy* 7 (1989):3–22.

Woodwell, G. "Biotic Effects on the Concentration of Atmospheric Carbon Dioxide: A Review and a Projection." In *Changing Climate*. Washington, D.C.: National Academy Press, 1983.

Woodwell, G., et al. "Global Deforestation: Contribution to Atmospheric Carbon Dioxide." *Science* 222 (1983):1081–1086.

INDEX

~~~~~~

ABOUT THE AUTHOR

Lynne Edgerton is a nationally and internationally recognized expert on global warming and sea level rise. As an attorney with the Natural Resources Defense Council in New York City for over six years, she was involved in many environmental protection battles to preserve the nation's coasts, most notably as a litigator in defense of unspoiled areas of Delaware Bay, and as a lobbyist and litigator to secure the phase-out of the ocean dumping of New York City's contaminated sewage sludge offshore of New York and New Jersey. Ms. Edgerton holds degrees from both the undergraduate and the law school of Vanderbilt University and from the Yale School of Law. She currently serves on the Board of the Climate Institute and is co-author of the coastal sections of the proposed International Convention on Climate Change submitted to the Intergovernmental Panel on Climate Change in February, 1990. She was one of a small group of Americans invited by the United Nations Environment Program and the World Meteorological Organization to attend the 1990 Second World Climate Conference in Geneva, Switzerland.

In 1989, Ms. Edgerton co-founded NRDC's Los Angeles office. Since her arrival in Southern California, she has focused on securing clean air and clean coasts. She is co-author of *No Safe Harbor*, an NRDC report on oil tanker safety which recommends ways to lessen oil spill risks. She is married to a physician and has two small children.

ALSO AVAILABLE FROM ISLAND PRESS

Ancient Forests of the Pacific Northwest
By Elliott A. Norse

Balancing on the Brink of Extinction: The Endangered Species Act and Lessons for the Future
Edited by Kathryn A. Kohm

Better Trout Habitat: A Guide to Stream Restoration and Management
By Christopher J. Hunter

The Challenge of Global Warming
Edited by Dean Edwin Abrahamson

Coastal Alert: Ecosystems, Energy, and Offshore Oil Drilling
By Dwight Holing

The Complete Guide to Environmental Careers
The CEIP Fund

Economics of Protected Areas
By John A. Dixon and Paul B. Sherman

Environmental Agenda for the Future
Edited by Robert Cahn

Environmental Disputes: Community Involvement in Conflict Resolution
By James E. Crowfoot and Julia M. Wondolleck

Fighting Toxics: A Manual for Protecting Your Family, Community, and Workplace
Edited by Gary Cohen and John O'Connor

Forests and Forestry in China: Changing Patterns of Resource Development
By S. D. Richardson

From *The Land*
Edited and compiled by Nancy P. Pittman

Hazardous Waste from Small Quantity Generators
By Seymour I. Schwartz and Wendy B. Pratt

Holistic Resource Management Workbook
By Allan Savory

In Praise of Nature
Edited and with essays by Stephanie Mills

138

The Living Ocean: Understanding and Protecting Marine Biodiversity
By Boyce Thorne-Miller and John Catena

Natural Resources for the 21st Century
Edited by R. Neil Sampson and Dwight Hair

The New York Environment Book
By Eric A. Goldstein and Mark A. Izeman

Overtapped Oasis: Reform or Revolution for Western Water
By Marc Reisner and Sarah Bates

Permaculture: A Practical Guide for a Sustainable Future
By Bill Mollison

Plastics: America's Packaging Dilemma
By Nancy A. Wolf and Ellen D. Feldman

The Poisoned Well: New Strategies for Groundwater Protection
Edited by Eric Jorgensen

Race to Save the Tropics: Ecology and Economics for a Sustainable Future
Edited by Robert Goodland

Recycling and Incineration: Evaluating the Choices
By Richard A. Denison and John Ruston

Reforming The Forest Service
By Randal O'Toole

The Rising Tide: Global Warming and World Sea Levels
By Lynne T. Edgerton

Rush to Burn: Solving America's Garbage Crisis?
From *Newsday*

Saving the Tropical Forests
By Judith Gradwohl and Russell Greenberg

War on Waste: Can America Win Its Battle With Garbage?
By Louis Blumberg and Robert Gottlieb

Western Water Made Simple
From *High Country News*

Wetland Creation and Restoration: The Status of the Science
Edited by Mary E. Kentula and Jon A. Kusler

Wildlife and Habitats in Managed Landscapes
Edited by Jon E. Rodiek and Eric G. Bolen

For a complete catalog of Island Press publications, please write:
Island Press, Box 7, Covelo, CA 95428. Or call 1–800–828–1302.

ISLAND PRESS
BOARD OF DIRECTORS

PETER R. STEIN, CHAIR
Managing Partner, Lyme Timber Company
Board Member, Land Trust Alliance

DRUMMOND PIKE, SECRETARY
Executive Director
The Tides Foundation

ROBERT E. BAENSCH
Director of Publishing
American Institute of Physics

PETER R. BORRELLI
Editor, *The Amicus Journal*
Natural Resources Defense Council

CATHERINE M. CONOVER

GEORGE T. FRAMPTON, JR.
President
The Wilderness Society

PAIGE K. MACDONALD
Executive Vice President/
Chief Operating Officer
World Wildlife Fund/The Conservation Foundation

HENRY REATH
President
Collectors Reprints, Inc.

CHARLES C. SAVITT
President
Center for Resource Economics/Island Press

SUSAN E. SECHLER
Director
Rural Economic Policy Program
Aspen Institute for Humanistic Studies

RICHARD TRUDELL
Executive Director
American Indian Lawyer Training Program